Lasers: Principles, Types and Applications

Lasers: Principles, Types and Applications

Gabriella Becker

NY Research Press
New York

Published by NY Research Press
118-35 Queens Blvd., Suite 400,
Forest Hills, NY 11375, USA
www.nyresearchpress.com

Lasers: Principles, Types and Applications
Gabriella Becker

International Standard Book Number: 978-1-63238-858-2 (Hardback)

Cataloging-in-Publication Data

Lasers : principles, types and applications / Gabriella Becker.
p. cm.
Includes bibliographical references and index.
ISBN 978-1-63238-858-2
1. Lasers. 2. Nonlinear optics. 3. Optical parametric oscillators.
4. Laser materials. I. Becker, Gabriella.
TA1675 .L37 2022
621.366--dc23

TABLE OF CONTENTS

PREFACE

This book aims to help a broader range of students by exploring a wide variety of significant topics related to this discipline. It will help students in achieving a higher level of understanding of the subject and excel in their respective fields. This book would not have been possible without the unwavered support of my senior professors who took out the time to provide me feedback and help me with the process. I would also like to thank my family for their patience and support.

A device that releases a beam of coherent light through optical amplification is called a laser. Laser stands for light amplification by stimulated emission of radiation. There are various components of a laser. It consists of gain medium, high reflector, laser pumping energy, laser beam and output coupler. The scientific study of theory and practice of lasers is known as laser science. Lasers have high temporal coherence. It allows them to emit light with a narrow spectrum. Laser beam profiler is used in measuring width, divergence and intensity profile of laser beams. Various types of lasers include gas lasers, semiconductor lasers, solid-state lasers, etc. Lasers are used in weapons, DNA sequencing instruments, printing, cutting and welding, lithography, small hand-held devices like laser pointers, etc. Laser medicine is the use of laser in medical diagnosis, treatments and therapies such as laser mammography, laser hair removal, laser surgery, etc. This book provides comprehensive insights into the field of laser science. Some of the diverse topics covered herein address the varied branches that fall under this category. This book is meant for students who are looking for an elaborate reference text on lasers.

A brief overview of the book contents is provided below:

Chapter - Laser and its Types

The device that produces light through the process of optical amplification is known as a laser. Some of the various types of lasers are gas discharge lasers, semiconductor diode lasers and optically pumped lasers. This chapter has been carefully written to provide an easy understanding of lasers, their construction as well as their types.

Chapter - Laser Physics

The branch of optics that describes the theories and practices of lasers is known as laser physics. Gain switching, mode-locking, Q-switching and laser pumping are some of the techniques used in laser physics. This chapter discusses in detail these techniques and methods related to laser physics.

Chapter – Gas Laser and its Types

A laser in which an electric current is discharged through a gas for producing coherent light is referred to as gas laser. Various types of gas laser include helium–neon lasers, chemical lasers, excimer lasers, ion lasers and metal-vapor lasers. The significant aspects of these types of gas lasers have been thoroughly discussed in this chapter.

Chapter – Solid State Lasers

Solid-state laser is a type of laser that makes use of gain medium which is a solid. A few types of solid state lasers are ruby laser, Nd: YAG laser, Er:YAG laser and diode-pumped solid-state laser. The topics elaborated in this chapter will help in gaining a better perspective about these types of solid-state lasers.

Chapter – Semiconductor Lasers

Semiconductor lasers consist of a semiconductor diode which produces an output beam that has the characteristics of laser light. The chapter closely examines various types of semiconductor lasers such as quantum cascade laser, quantum dot laser, transistor laser, interband cascade laser, quantum well laser, etc.

Chapter – Applications in Diverse Fields

Lasers are used in various fields such as defense, surgery, medicine and in military operations. The industrial and commercial applications of lasers include laser cutting, laser pointers, laser drilling, laser engraving and laser beam welding. This chapter has been carefully written to provide an easy understanding of these applications of lasers.

Gabriella Becker

Laser and its Types

1

- **Laser Construction**
- **Types of Lasers**

The device that produces light through the process of optical amplification is known as a laser. Some of the various types of lasers are gas discharge lasers, semiconductor diode lasers and optically pumped lasers. This chapter has been carefully written to provide an easy understanding of lasers, their construction as well as their types.

Laser is a device that stimulates atoms or molecules to emit light at particular wavelengths and amplifies that light, typically producing a very narrow beam of radiation. The emission generally covers an extremely limited range of visible, infrared, or ultraviolet wavelengths. Many different types of lasers have been developed, with highly varied characteristics. *Laser* is an acronym for "light amplification by the stimulated emission of radiation."

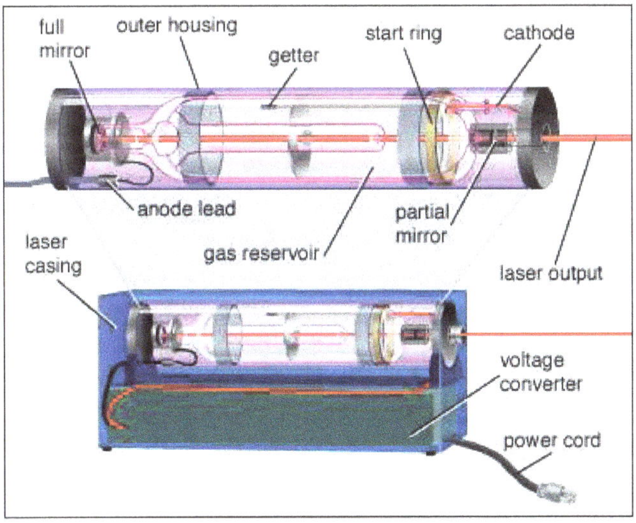

The laser is an outgrowth of a suggestion made by Albert Einstein in 1916 that under the proper circumstances atoms could release excess energy as light—either spontaneously or when stimulated by light. German physicist Rudolf Walther Ladenburg first observed stimulated emission in 1928, although at the time it seemed to have no practical use.

In 1951 Charles H. Townes, then at Columbia Universityin New York City, thought of a way to generate stimulated emission at microwave frequencies. At the end of 1953, he demonstrated a working device that focused "excited" ammonia molecules in a resonant microwave cavity, where they emitted a pure microwave frequency. Townes named the device a maser, for "microwave amplification by the stimulated emission of radiation." Aleksandr Mikhaylovich Prokhorov and Nikolay Gennadiyevich Basov of the P.N. Lebedev Physical Institute in Moscow independently described the theory of maser operation. For their work all three shared the 1964 Nobel Prize for Physics.

An intense burst of maser research followed in the mid-1950s, but masers found only a limited range of applications as low-noise microwave amplifiers and atomic clocks. In 1957 Townes proposed to his brother-in-law and former postdoctoral student at Columbia University, Arthur L. Schawlow (then at Bell Laboratories), that they try to extend maser action to the much shorter wavelengths of infrared or visible light. Townes also had discussions with a graduate student at Columbia University, Gordon Gould, who quickly developed his own laser ideas. Meanwhile, Gould coined the word *laser* and wrote a patent application. Whether Townes or Gould should be credited as the "inventor" of the laser thus became a matter of intense debate and led to years of litigation. Eventually, Gould received a series of four patents starting in 1977 that earned him millions of dollars in royalties.

The Townes-Schawlow proposal led several groups to try building a laser. The Gould proposal became the basis of a classified military contract. Success came first to Theodore H. Maiman, who took a different approach at Hughes Research Laboratories in Malibu, California. He fired bright pulses from a photographer's flash lamp to excite chromium atoms in a crystal of synthetic ruby, a material he chose because he had studied carefully how it absorbed and emitted light and calculated that it should work as a laser. On May 16, 1960, he produced red pulses from a ruby rod about the size of a fingertip. In December 1960 Ali Javan, William Bennett, Jr., and Donald Herriott at Bell Labs built the first gas laser, which generated a continuous infrared beam from a mixture of helium and neon. In 1962 Robert N. Hall and coworkers at the General Electric Research and Development Center in Schenectady, New York, made the first semiconductor laser.

While lasers quickly caught the public imagination, perhaps for their similarity to the "heat rays" of science fiction, practical applications took years to develop. A young physicist named Irnee D'Haenens, while working with Maiman on the ruby laser, joked that the device was "a solution looking for a problem," and the line lingered in the laser community for many years. Townes and Schawlow had expected laser beams to be used in basic research and to send signals through air or space. Gould envisioned more powerful beams capable of cutting and drilling many materials. A key early success came in late 1963 when two researchers at the University of Michigan, Emmett Leith and Juris Upatnieks, used lasers to make the first three-dimensional holograms.

Helium-neon lasers were the first lasers with broad commercial applications. Because they could be adjusted to generate a visible red beam instead of an infrared beam, they found immediate use projecting straight lines for alignment, surveying, construction, and irrigation. Soon eye surgeons were using pulses from ruby lasers to weld detached retinas back in place without cutting into the eye. The first large-scale application for lasers was the laser scanner for automated checkout in supermarkets, which was developed in the mid-1970s and became common a few years later. Compact disc audio players and laser printers for personal computers soon followed.

Lasers have become standard tools in diverse applications. Laser pointers highlight presentation points in lecture halls, and laser target designators guide smart bombs to their targets. Lasers weld razor blades, write patterns on objects on production lines without touching them, remove unwanted hair, and bleach tattoos. Laser rangefinders in space probes profiled the surfaces of Mars and the asteroid Eros in unprecedented detail. In the laboratory, lasers have helped physicists to cool atoms to within a tiny fraction of a degree of absolute zero.

Fundamental Principles

Energy Levels and Stimulated Emissions

Laser emission is shaped by the rules of quantum mechanics, which limit atoms and molecules to having discrete amounts of stored energy that depend on the nature of the atom or molecule. The lowest energy level for an individual atom occurs when its electrons are all in the nearest possible orbits to its nucleus. This condition is called the ground state. When one or more of an atom's electrons have absorbed energy, they can move to outer orbits, and the atom is then referred to as being "excited." Excited states are generally not stable; as electrons drop from higher-energy to lower-energy levels, they emit the extra energy as light.

Einstein recognized that this emission could be produced in two ways. Usually, discrete packets of light known as photons are emitted spontaneously, without outside intervention. Alternatively, a passing photon could stimulate an atom or molecule to emit light—if the passing photon's energy exactly matched the energy that an electron would release spontaneously when dropping to a lower-energy configuration. Which process dominates depends on the ratio of lower-energy to higher-energy configurations. Ordinarily, lower-energy configurations predominate. This means that a spontaneously emitted photon is more likely to be absorbed and raise an electron from a lower-energy configuration to a higher-energy configuration than to stimulate a higher-energy configuration to drop to a lower-energy configuration by emitting a second photon. As long as lower-energy states are more common, stimulated emission will die out.

However, if higher-energy configurations predominate (a condition known as population inversion), spontaneously emitted photons are more likely to stimulate further emissions, generating a cascade of photons. Heat alone does not produce a population inversion; some process must selectively excite the atoms or molecules. Typically, this is done by illuminating the laser material with bright light or by passing an electric current through it.

The simplest conceivable system, such as the ammonia maser built by Townes, has only two energy levels. More useful laser systems involve three or four energy levels. In a three-level laser, the material is first excited to a short-lived high-energy state that spontaneously drops to a somewhat lower-energy state with an unusually long lifetime, called a metastable state. The metastable state is important because it traps and holds the excitation energy, building up a population inversion that can be further stimulated to emit radiation, dropping the species back to the ground state. The ruby laser developed by Theodore Maiman is an example of a three-level laser.

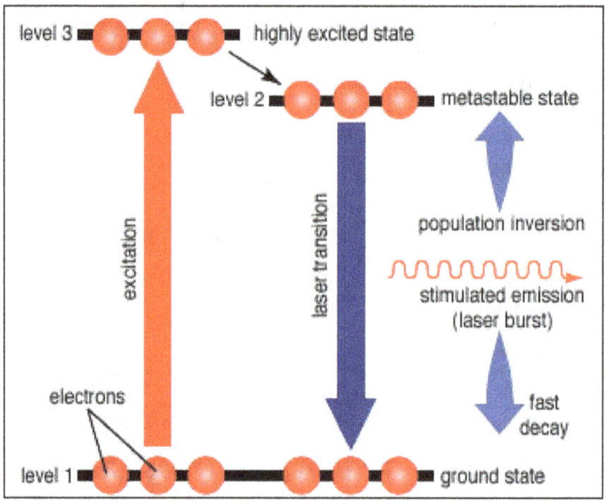

Three-level laser A burst of energy excites electrons in more than half of the atoms from their ground state to a higher state, creating a population inversion.

The electrons then drop into a long-lived state with slightly less energy, where they can be stimulated to quickly shed excess energy as a laser burst, returning the electrons to a stable ground state.

Unfortunately, the three-level laser works only if the ground state is depopulated. As atoms or molecules emit light, they accumulate in the ground state, where they can absorb the stimulated emission and shut down laser action, so most three-level lasers can only generate pulses. This difficulty is overcome in the four-level laser, where an extra transition state is located between metastable and ground states. This allows many four-level lasers to emit a steady beam for days on end.

As in a three-level laser, the atoms first drop to a long-lived metastable state where they can be stimulated to emit excess energy. However, instead of dropping to the ground state, they stop at another state above the ground state from which they can more easily be excited back up to the higher metastable state, thereby maintaining the population inversion needed for continuous laser operation.

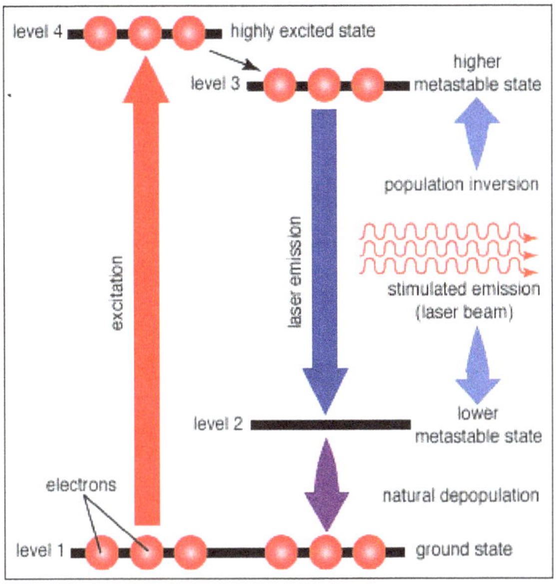

Four-level laserA sustained laser beam can be achieved by using atoms that have two relatively stable levels between their ground state and a higher-energy excited state.

Laser Elements

Population inversions can be produced in a gas, liquid, or solid, but most laser media are gases or solids. Typically, laser gases are contained in cylindrical tubes and excited by an electric current or external light source, which is said to "pump" the laser. Similarly, solid-state lasers may use semiconductors or transparent crystals with small concentrations of light-emitting atoms.

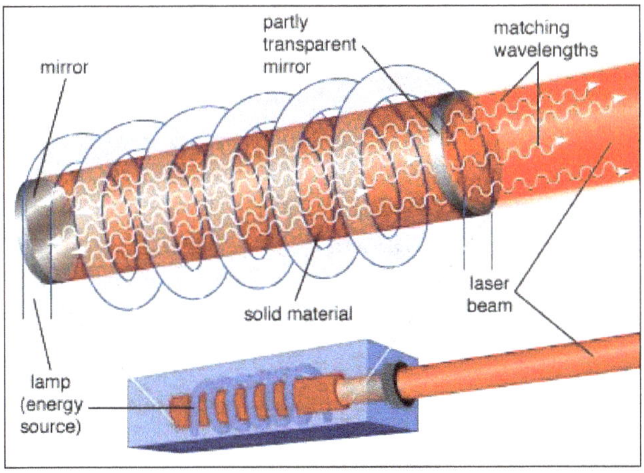

Laser producing a beam.

An optical resonator is needed to build up the light energy in the beam. The resonator is formed by placing a pair of mirrors facing each other so that light emitted along the line between the mirrors is reflected back and forth. When a population inversion is created in the medium, light reflected back and forth increases in intensity with each pass through the laser medium. Other light leaks around the mirrors without being amplified. In an actual laser cavity, one or both mirrors transmit a fraction of the incident light. The fraction of light transmitted—that is, the laser beam—depends on the type of laser. If the laser generates a continuous beam, the amount of light added by stimulated emission on each round trip between the mirrors equals the light emerging in the beam plus losses within the optical resonator.

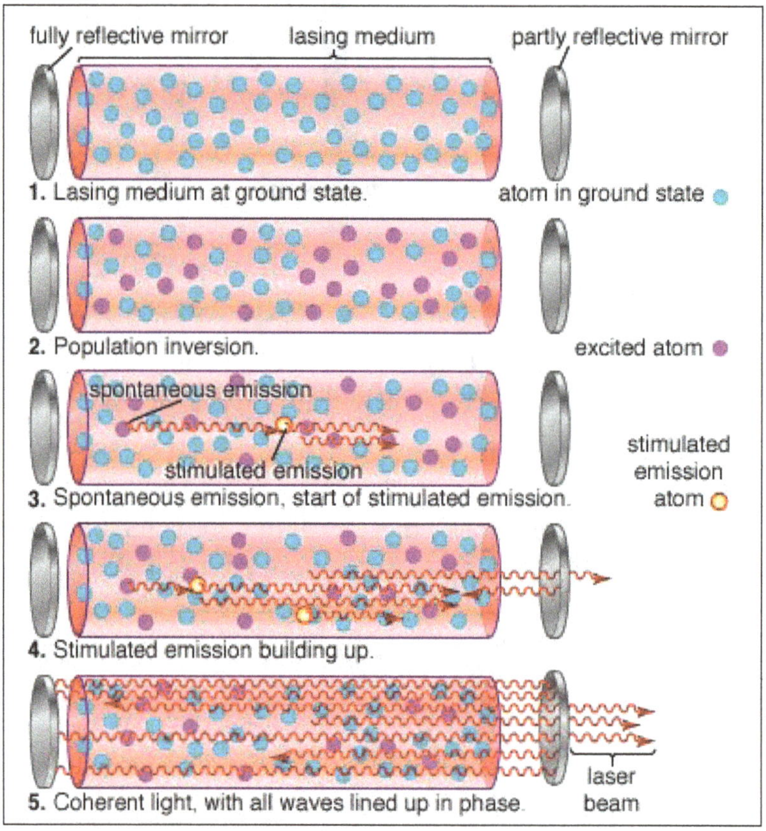

Stimulated emission in a laser cavity.

The combination of laser medium and resonant cavity forms what often is called simply a laser but technically is a laser oscillator. Oscillation determines many laser properties, and it means that the device generates light internally. Without mirrors and a resonant cavity, a laser would just be an optical amplifier, which can amplify light from an external source but not generate a beam internally. Elias Snitzer, a researcher at American Optical, demonstrated the first optical amplifier in 1961, but such devices were little used until the spread of communications based on fibre optics.

Laser Beam Characteristics

Laser light generally differs from other light in being focused in a narrow beam, limited to a narrow range of wavelengths (often called "monochromatic"), and consisting of waves that are in phase with each other. These properties arise from interactions between the process of stimulated emission, the resonant cavity, and the laser medium.

Stimulated emission produces a second photon identical to the one that stimulated the emission, so the new photon has the same phase, wavelength, and direction—that is, the two are coherent with respect to each other, with peaks and valleys in phase. Both the original and the new photon can then stimulate the emission of other identical photons. Passing the light back and forth through a resonant cavity enhances this uniformity, with the degree of coherence and the narrowness of the beam depending on the laser design.

Although a visible laser produces what looks like a point of light on the opposite wall of a room, the alignment, or collimation, of the beam is not perfect. The extent of beam spreading depends on both the distance between the laser mirrors and diffraction, which scatters light at the edge of an aperture. Diffraction is proportional to the laser wavelength divided by the size of the emitting aperture; the larger the aperture is, the more slowly the beam spreads. A red helium-neon laser emits from a one-millimetre aperture at a wavelength of 0.633 micrometre, generating a beam that diverges at an angle of about 0.057 degree, or one milliradian. Such a small angle of divergence will produce a one-metre spot at a distance of one kilometre. In contrast, a typical flashlight beam produces a similar one-metre spot within a few metres. Not all lasers produce tight beams, however. Semiconductor lasers emit light near one micrometre wavelength from an aperture of comparable size, so their divergence is 20 degrees or more, and external optics are needed to focus their beams.

The output wavelength depends on the laser material, the process of stimulated emission, and the optics of the laser resonator. For each transition between energy levels, a material can support stimulated emission over a limited range of wavelengths; the extent of that range varies with the nature of the material and the transition. The probability of stimulated emission varies with wavelength, and the process concentrates emission at wavelengths where that probability is the highest.

Resonant cavities support laser oscillation at wavelengths that meet a resonant condition—an integral number N of wavelengths λ must equal the distance light travels during a round trip between the mirrors. If the cavity length is L and the refractive index of the material in the laser cavity is n, the round-trip distance $2L$ must equal $N\lambda/n$, or $2L = N\lambda/n$. Each resonance is called a longitudinal mode. Except in semiconductor lasers, cavities are thousands of wavelengths long, so the wavelengths of adjacent modes are closely spaced—and usually the laser simultaneously emits light on two or more wavelengths within 0.1 percent of each other. These beams are monochromatic for most practical applications; other optics can be added to limit laser oscillation to a single longitudinal mode and an even narrower range of wavelengths. The best laboratory lasers emit a range of wavelengths that differ by less than 0.0000001 percent.

The narrower the range of wavelengths, the more coherent the beam—meaning the more precisely every light wave in the beam is in exact synchronization with every other one. This is measured by a quantity called coherence length. If the centre of the range of wavelengths emitted is λ and the range of wavelengths emitted is $\Delta\lambda$, this coherence length equals $\lambda^2=2\Delta\lambda$. Typical coherence lengths range from millimetres to metres. Such long coherence lengths are essential, for instance, to record holograms of three-dimensional objects.

Lasers can generate pulsed or continuous beams, with average powers ranging from microwatts to over a million watts in the most powerful experimental lasers. A laser is called continuous-wave if its output is nominally constant over an interval of seconds or longer; one example is the steady red beam from a laser pointer. Pulsed lasers concentrate their output energy into brief high-power bursts. These lasers can fire single pulses or a series of pulses at regular intervals. Instantaneous power can be extremely high at the peak of a very short pulse. Laboratory lasers have generated peak power exceeding 10^{15} watts for intervals of about 10^{-12} second.

Pulses can be compressed to extremely short duration, about 5 femtoseconds (5×10^{-15} second) in laboratory experiments, in order to "freeze" the action during events that occur very rapidly, such as stages in chemical reactions. Laser pulses also can be focused to concentrate high powers on small spots, much as a magnifier focuses sunlight onto a small spot to ignite a piece of paper.

Types of Lasers

Crystals, glasses, semiconductors, gases, liquids, beams of high-energy electrons, and even gelatin doped with suitable materials can generate laser beams. In nature, hot gases near bright stars can generate strong stimulated emission at microwave frequencies, although these gas clouds lack resonant cavities, so they do not produce beams.

In crystal and glass lasers, such as Maiman's first ruby laser, light from an external source excites atoms, known as dopants, that have been added to a host material at low concentrations. Important examples include glasses and crystals doped with the rare-earth element neodymium and glasses doped with erbium or ytterbium, which can be drawn into fibres for use as fibre-optic lasers or amplifiers. Titaniumatoms doped into synthetic sapphire can generate stimulated emission across an exceptionally broad range and are used in wavelength-tunable lasers.

Many different gases can function as laser media. The common helium-neon laser contains a small amount of neon and a much larger amount of helium. The helium atoms capture energy from electrons passing through the gas and transfer it to the neon atoms, which emit light. The best-known helium-neon lasers emit red light, but they also can be made to emit yellow, orange, green, or infrared light; typical powers are in the milliwatt range. Argon and krypton atoms that have been stripped of one or two electrons can generate milliwatts to watts of laser light at visible and ultraviolet wavelengths. The most powerful commercial gas laser is the carbon-dioxide laser, which can generate kilowatts of continuous power.

The most widely used lasers today are semiconductor diode lasers, which emit visible or infrared light when an electric current passes through them. The emission occurs at the interface between two regions doped with different materials. The *p-n* junction can act as a laser medium, generating stimulated emission and providing lasing action if it is inside a suitable cavity. Conventional edge-emitting semiconductor lasers have mirrors on opposite edges of the *p-n* junction, so light oscillates in the junction plane. Vertical-cavity surface-emitting lasers (VCSELs) have mirrors above and below the *p-n* junction, so light resonates perpendicular to the junction. The wavelength depends on the semiconductor compound.

A few other types of lasers are used in research. In dye lasers the laser medium is a liquid containing organic dye molecules that can emit light over a range of wavelengths; adjusting the laser cavity changes, or tunes, the output wavelength. Chemical lasers are gas lasers in which a chemical reactiongenerates the excited molecules that produce stimulated emission. In free-electron lasers stimulated emission comes from electrons passing through a magnetic field that periodically varies in direction and intensity, causing the electrons to accelerate and release light energy. Because the electrons do not transition between well-defined energy levels, some specialists question whether a free-electron laser should be called a laser, but the label has stuck. Depending on the energy of the electron beam and variations in the magnetic field, free-electron lasers can be tuned across a wide range of wavelengths. Both free-electron and chemical lasers can emit high powers.

Laser Applications

Lasers deliver coherent, monochromatic, well-controlled, and precisely directed light beams. Although lasers make poor choices for general-purpose illumination, they are ideal for concentrating light in space, time, or particular wavelengths. For example, many people were first introduced to lasers by concerts in the early 1970s that incorporated laser light shows, in which moving laser beams of different colours projected changing patterns on planetarium domes, concert-hall ceilings, or outdoor clouds.

Most laser applications fall into one of a few broad categories: (1) transmission and processing of information, (2) precise delivery of energy, and (3) alignment, measurement, and imaging. These categories cover diverse applications, from pinpoint energy delivery for delicate surgery to heavy-duty welding and from the mundane alignment of suspended ceilings to laboratory measurements of atomic properties.

Transmission and Processing of Information

Laser Scanners

The ability to focus laser beams onto very small spots and to switch them on and off billions of times per second makes lasers important tools in telecommunications and information processing. In laser supermarket scanners, a rotating mirror scans a red beam while clerks move packages across the beam. Optical sensors detect light reflected from striped bar codes on packages, decode the symbol, and relay the information to a computer so that it can add the price to the bill.

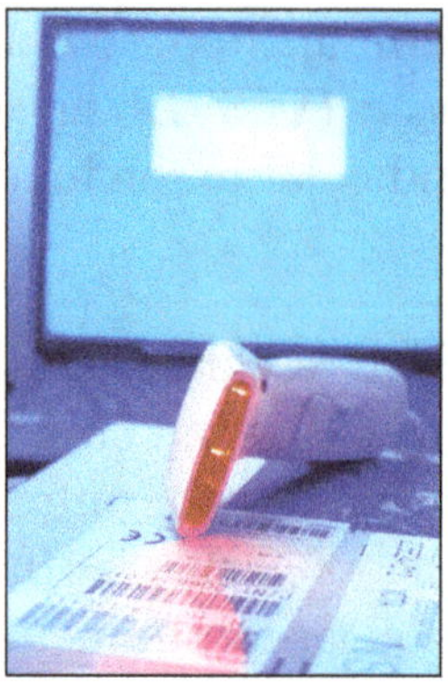

Since their introduction in 1974, laser scanners for reading universal product codes (UPC), or bar codes, have become common in retail stores.

Optical Discs

Tiny, inexpensive semiconductor lasers read data from a growing variety of optical compact disc formats to play music, display video recordings, and read computer software. Audio compact discs, using infrared lasers, were introduced around 1980; CD-ROMs (compact disc read-only memory) for computer data soon followed. Newer optical drives use more powerful lasers to record data on light-sensitive discs called CD-R (recordable) or CD-RW (read/write), which can be played in ordinary CD-ROM drives. DVDs (digital video, or versatile, discs) work similarly, but they use a shorter-wavelength red laser to read smaller spots, so the discs can hold enough information to play a digitized motion picture. A new generation of discs called Blu-ray uses blue-light lasers to read and store data at an even higher density.

Fibre-optic Communication Systems

Fibre-optic communication systems that transmit signals more than a few kilometres also use semiconductor laser beams. The optical signals are sent at infrared wavelengths of 1.3 to 1.6 micrometres, where glass fibres are most transparent. This technology has become the backbone of the global telecommunications network, and most telephone calls traveling beyond the confines of a single town go part of the way through optical fibres.

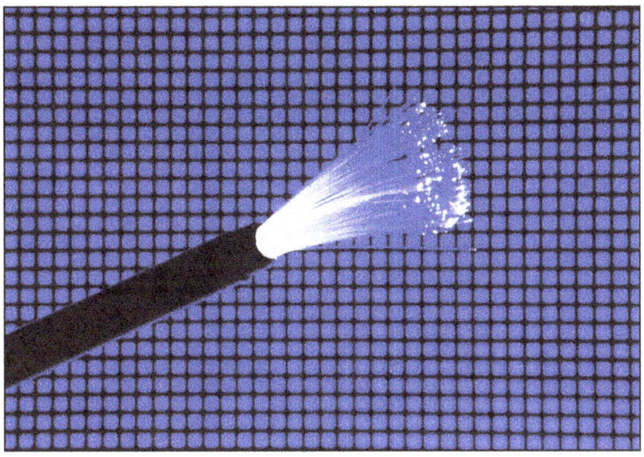

Modern communication systems use fibre optic cables, which may have as many as a thousand individual fibres, because of a variety of benefits, such as greater data capacity, immunity to electro-magnetic interference, no risk of starting electrical fires, and improved security of communications.

Precise Delivery of Energy

Industrial Uses

Laser energy can be focused in space and concentrated in time so that it heats, burns away, or vaporizes many materials. Although the total energy in a laser beam may be small, the concentrated power on small spots or during short intervals can be enormous. Although lasers cost much more than mechanical drills or blades, their different properties allow them to perform otherwise difficult tasks. A laser beam does not deform flexible materials as a mechanical drill would, so it can drill holes in materials such as soft rubber nipples for baby bottles. Likewise, laser beams can drill or cut into extremely hard materials without dulling bits or blades. For example, lasers have drilled holes in diamond dies used for drawing wire.

Medical Applications

Surgical removal of tissue with a laser is a physical process similar to industrial laser drilling. Carbon-dioxide lasers burn away tissue because their infrared beams are strongly absorbed by the water that makes up the bulk of living cells. A laser beam cauterizes the cuts, stopping bleeding in blood-rich tissues such as the female reproductive tract or the gums. Laser wavelengths near one micrometre can penetrate the eye, welding a detached retina back into place, or cutting internal membranes that often grow cloudy after cataract surgery. Less-intense laser pulses can destroy abnormal blood vessels that spread across the retina in patients suffering from diabetes, delaying the blindness often associated with the disease. Ophthalmologists surgically correct visual defects by removing tissue from the cornea, reshaping the transparent outer layer of the eye with intense ultraviolet pulses.

Photodynamic therapy (PDT) fibre optic surgery: A photosensitive drug absorbed by cancer cells can be activated by a laser beam guided through optical fibres to selectively destroy a tumour.

Through the use of optical fibres similar to the tiny strands of glass that carry information in telephone systems, laser light can be delivered to places within the body that the beams could not otherwise reach. One important example involves threading a fibre through the urethra and into the kidney so that the end of the fibre can deliver intense laser pulses to kidney stones. The laser energy splits the stones into fragments small enough to pass through the urethra without requiring surgical incisions. Fibres also can be inserted through small incisions to deliver laser energy to precise spots in the knee joint during arthroscopic surgery.

Another medical application for lasers is in the treatment of skin conditions. Pulsed lasers can bleach certain types of tattoos as well as dark-red birthmarks called portwine stains. Cosmetic laser treatments include removing unwanted body hair and wrinkles.

High-energy Lasers

Scientists have shown that lasers can concentrate extremely high powers in either pulses or continuous beams. Major applications for these high-power levels are fusion research, nuclear weapons testing, and missile defense.

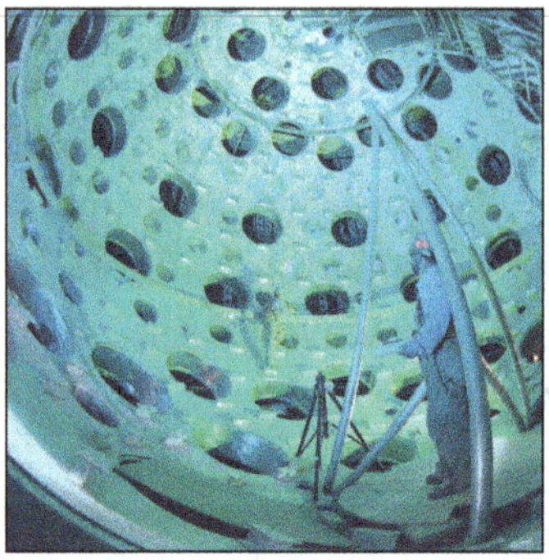

Laser-activated fusionInterior of the U.S. Department of Energy's National Ignition Facility (NIF), located at Lawrence Livermore National Laboratory, Livermore, California. The NIF target chamber uses a high-energy laser to heat fusion fuel to temperatures sufficient for thermonuclear ignition. The facility is used for basic science, fusion energy research, and nuclear weapons testing.

Extremely high temperatures and pressures are needed to force atomic nuclei to fuse together, releasing energy. In the 1960s physicists at the Lawrence Livermore National Laboratory in California calculated that intense laser pulses could produce those conditions by heating and compressing tiny pellets containing mixtures of hydrogen isotopes. They suggested using these "microimplosions" both to generate energy for civilian use and to simulate the implosion of a hydrogen bomb, which involves similar processes. Since then, Livermore has built a series of lasers to test and refine these theories, primarily for the U.S. government's nuclear weapons program.

Military laser weapon research also dates back to the 1960s, but it attracted little attention until President Ronald Reagan launched the Strategic Defense Initiative in 1983. High-energy lasers offer a way to deliver destructive energy to targets at the speed of light, which is very attractive for fast-moving targets such as nuclear missiles. Military researchers have tested high-energy lasers for use as weapons on land, at sea, in the air, and in space, although no high-energy lasers have been placed in orbit. Experiments have shown that massive lasers can generate high powers; however, tests have also shown that the atmosphere distorts such powerful beams, causing them to spread out and miss their targets. These problems and the end of the Cold War slowed research on laser weapons, though interest continues in laser weapons to defend against smaller-scale missile attacks.

Alignment, Measurement and Imaging

Surveying

Surveyors and construction workers use laser beams to draw straight lines through the air. The beam itself is not visible in the air except where scattered by dust or haze, but it projects a bright point on a distant object. Surveyors bounce the beam off a mirror to measure direction and angle. The beam can set an angle for grading irrigated land, and a rotating beam can define a smooth plane for construction workers installing walls or ceilings.

Pulsed laser radar can measure distance in the same manner as microwave radar by timing how long it takes a laser pulse to bounce back from a distant object. For instance, in 1969 laser radar precisely measured the distance from the Earth to the Moon, and in the 1970s military laser range finders were developed to measure the distance to battlefield targets accurately. Laser range finding is now widely used for remote sensing. Instruments flown on aircraft can profile the layers of foliage in a forest, and the Mars Global Surveyor used a laser altimeter to map elevations on the Martian surface.

The image clearly shows the shield volcano's relative flatness and gently sloping profile, the steep outward-facing cliff at its base (buried in places under lava that has flowed into the surrounding plains), and the complex caldera of intersecting craters at the summit.

Olympus Mons, the highest point on Mars, in a computer-generated oblique view made by combining photos obtained by the Viking mission in the 1970s with topographic data gathered by Mars Global Surveyor a quarter century later.

Interferometry and Holography

The coherence of laser light is crucial for interferometry and holography, which depend on interactions between light waves to make extremely precise measurements and to record three-dimensional images. The result of adding light waves together depends on their relative phases. If the peaks of one align with the valleys of the other, they will interfere destructively to cancel each other out; if their peaks align, they will interfere constructively to produce a bright spot. This effect can be used for measurement by splitting a beam into two identical halves that follow different paths. Changing one path just half a wavelength from the other will shift the two out of phase, producing a dark spot. This technique has proved invaluable for precise measurements of very small distances.

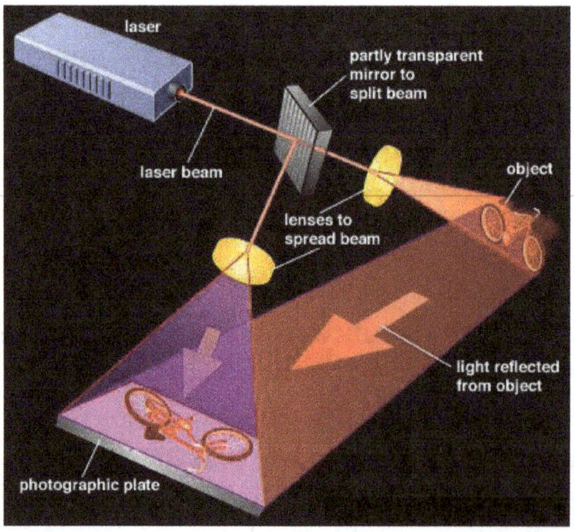

Holography uses no camera. Instead, two beams of light from a single laser shine on a piece of film. One of the beams reflects from the object.

Holograms are made by splitting a laser beam into two identical halves, using one beam to illuminate an object. This object beam then is combined with the other half—the reference beam—in the plane of a photographic plate, producing a random-looking pattern of light and dark zones that record the wave front of light from the object. Later, when laser light illuminates that pattern from the same angle as the reference beam, it is scattered to reconstruct an identical wave front of light, which appears to the viewer as a three-dimensional image of the object. Holograms now can be mass-produced by an embossing process, as used on credit cards, and do not have to be viewed in laser light.

Research Tool

The ability to control laser wavelength and pulse duration precisely has proved invaluable for fundamental research in physics and other sciences. Lasers have been particularly important in spectroscopy, the study of the light absorbed and emitted when atoms and molecules make transitions between energy levels, which can reveal the inner workings of atoms. Lasers can concentrate much more power into a narrow range of wavelengths than other light sources, which makes them invaluable in analyzing fine spectroscopic details.

For example, simultaneously illuminating samples with laser beams coming from opposite directions can cancel the effects of the random motions of atoms or molecules in a gas. This technique has greatly improved the precision of the measurement of the Rydberg constant, which is critical in calculations of atomic properties, and it earned Arthur Schawlow a share of the 1981 Nobel Prize for Physics. Nicolaas Bloembergen shared the prize for developing other types of high-precision laser spectroscopy.

Since that early work, laser spectroscopy has expanded considerably. Laser pulses have been used to take snapshots of chemical reactions as they occur, on time scales faster than atomic vibrations in a molecule. These techniques have given chemists new ways to understand chemical physics, and they earned Ahmed Zewail the 1999 Nobel Prize for Chemistry. Thanks to his work, the Nobel Committee wrote, "we have reached the end of the road: no chemical reactions take place faster than this."

Physicists also have used the subtle forces exerted by laser beams to slow and trap atoms, molecules, and small particles. Arthur Ashkin, a researcher at Bell Labs, showed that a tightly focused horizontal laser beam could trap atoms in the zone with highest light intensity, a technique called "optical tweezers" now used in a variety of research. Other research has shown that laser illumination can slow the motion of atoms if its wavelength is tuned to a point slightly off the wavelength of peak absorption. The atoms repeatedly absorb photons from the beam and then emit photons in random directions. The photon momentum slows the motion toward the laser beam. Placing the atoms at the junction of six laser beams aimed at right angles to each other slows their momentum in all directions, producing a clump of atoms less than 0.001 degree above absolute zero. Adding a magnetic field improves confinement and can reduce their temperature to less than one-millionth degree above absolute zero. These techniques have led to the creation of a new state of matter, called a Bose-Einstein condensate, and they earned Steven Chu, Claude Cohen-Tannoudji, and William D. Phillips the 1997 Nobel Prize for Physics.

Principle of Lasers

The Bohr Atom

In 1915, Neils Bohr proposed a model of the atom that explained a wide variety of phenomena that were puzzling scientists in the late 19th century. This simple model became the basis for the field of quantum mechanics and, although not fully accurate by today's understanding, still is useful for demonstrating laser principles. In Bohr's model, shown in figure, electrons orbit the nucleus of an atom. Unlike earlier "planetary" models, the Bohr atom has a limited number of fixed orbits that are available to the electrons. Under the right circumstances an electron can go from its ground state (lowest-energy orbit) to a higher (excited) state, or it can decay from a higher state to a lower state, but it cannot remain between these states. The allowed energy states are called "quantum" states and are referred to by the principal "quantum numbers" 1, 2, 3, etc. The quantum states are represented by an energy-level diagram.

For an electron to jump to a higher quantum state, the atom must receive energy from the outside world. This can happen through a variety of mechanisms such as inelastic or semielastic collisions with other atoms and absorption of energy in the form of electromagnetic radiation (e.g., light). Likewise, when an electron drops from a higher state to a lower state, the atom must give off energy, either as kinetic activity (nonradiative transitions) or as electromagnetic radiation (radiative transitions).

Photons and Energy

In the 1600s and 1700s, early in the modern study of light, there was a great controversy about light's nature. Some thought that light was made up of particles, while others thought that it was made up of waves. Both concepts explained some of the behavior of light, but not all. It was finally determined that light is made up of particles called "photons" which exhibit both particle-like and wave-like properties. Each photon has an intrinsic energy determined by the equation,

$$E = hv$$

Where, v is the frequency of the light and h is Planck's constant. Since, for a wave, the frequency and wavelength are related by the equation,

$$\lambda v = c$$

Where, v is the wavelength of the light and c is the speed of light in a vacuum, equation ($E = hv$) can be rewritten as,

$$E = \frac{hc}{\lambda}.$$

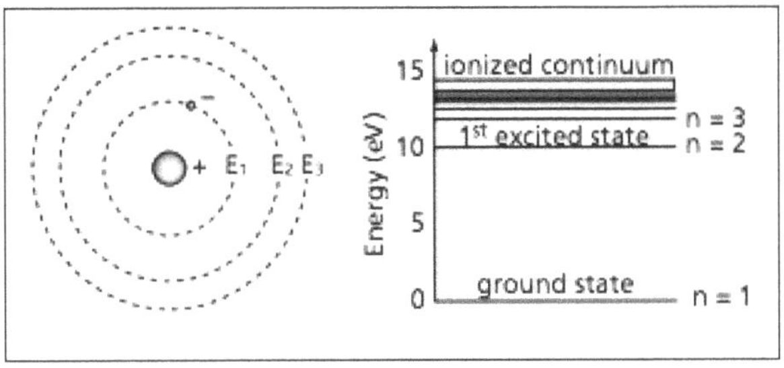

The Bohr atom and a simple energy-level diagram.

It is evident from this equation that the longer the wavelength of the light, the lower the energy of the photon; consequently, ultraviolet light is much more "energetic" than infrared light. Returning to the Bohr atom: for an atom to absorb light (i.e., for the light energy to cause an electron to move from a lower energy state E_n to a higher energy state E_m), the energy of a single photon must equal, almost exactly, the energy difference between the two states. Too much energy or too little energy and the photon will not be absorbed. Consequently, the wavelength of that photon must be,

$$\lambda = \frac{hc}{\Delta E}$$

Where,

$$\Delta E = E_m - E_n.$$

Likewise, when an electron decays to a lower energy level in a radiative transition, the photon of light given off by the atom must also have an energy equal to the energy difference between the two states.

Spontaneous and Stimulated Emission

In general, when an electron is in an excited energy state, it must eventually decay to a lower level, giving off a photon of radiation. This event is called "spontaneous emission," and the photon is emitted in a random direction and a random phase. The average time it takes for the electron to decay is called the time constant for spontaneous emission, and is represented by τ.

On the other hand, if an electron is in energy state E_2, and its decay path is to E_1, but, before it has a chance to spontaneously decay, a photon happens to pass by whose energy is approximately $E_2 - E_1$, there is a probability that the passing photon will cause the electron to decay in such a manner that a photon is emitted at exactly the same wavelength, in exactly the same direction, and with exactly the same phase as the passing photon. This process is called "stimulated emission." Absorption, spontaneous emission, and stimulated emission are illustrated in figure.

Now consider the group of atoms shown in figure: all begin in exactly the same excited state, and most are effectively within the stimulation range of a passing photon. We also will assume that τ is very long, and that the probability for stimulated emission is 100 percent. The incoming (stimulating) photon interacts with the first atom, causing stimulated emission of a coherent photon; these two photons then interact with the next two atoms in line, and the result is four coherent photons, on down the line. At the end of the process, we will have eleven coherent photons, all with identical phases and all traveling in the same direction. In other words, the initial photon has been "amplified" by a factor of eleven.

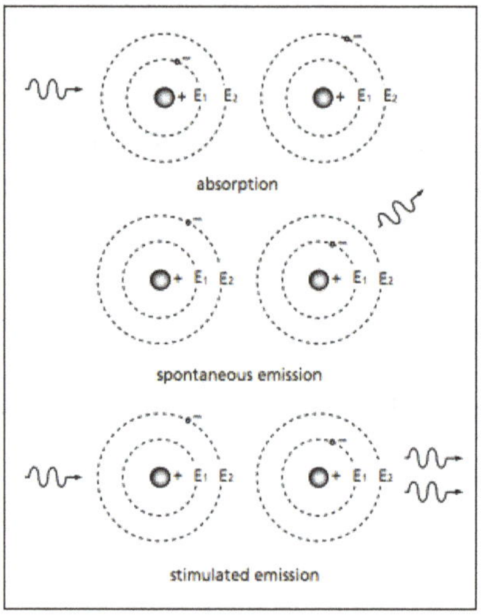

Spontaneous and stimulated emission.

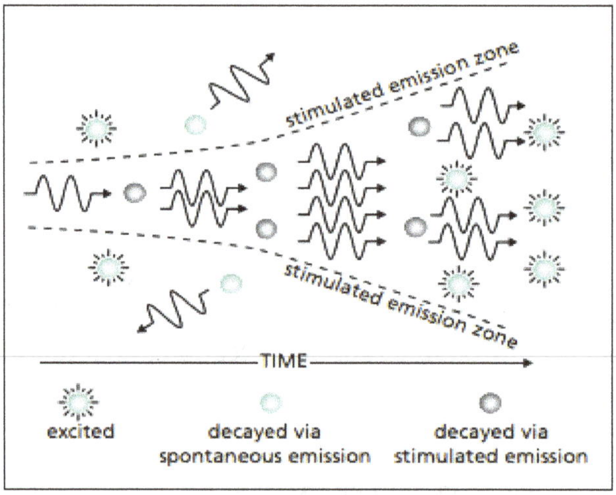

Amplification by stimulated emission.

Of course, in any real population of atoms, the probability for stimulated emission is quite small. Furthermore, not all of the atoms are usually in an excited state; in fact, the opposite is true. Boltzmann's principle, a fundamental law of thermodynamics, states that, when a collection of atoms is at thermal equilibrium, the relative population of any two energy levels is given by,

$$\frac{N_2}{N_1} = \exp\left(-\frac{E_2 - E_1}{kT}\right)$$

where N_2 and N_1 are the populations of the upper and lower energy states, respectively, T is the equilibrium temperature, and k is Boltzmann's constant. Substituting hn for E_2-E_1 yields.

$$\Delta N \equiv N_1 - N_2 \equiv (1 - e^{-hv/kt})N_1.$$

For a normal population of atoms, there will always be more atoms in the lower energy levels than in the upper ones. Since the probability for an individual atom to absorb a photon is the same as the probability for an excited atom to emit a photon via stimulated emission, the collection of real atoms will be a net absorber, not a net emitter, and amplification will not be possible. Consequently, to make a laser, we have to create a "population inversion."

Population Inversion

Atomic energy states are much more complex than indicated by the description above. There are many more energy levels, and each one has its own time constants for decay. The four-level energy diagram shown in figure is representative of some real lasers. The electron is pumped (excited) into an upper level E_4 by some mechanism (for example, a collision with another atom or absorption of high-energy radiation). It then decays to E_3, then to E_2, and finally to the ground state E_1. Let us assume that the time it takes to decay from E_2 to E_1 is much longer than the time it takes to decay from E_2 to E_1. In a large population of such atoms, at equilibrium and with a continuous pumping process, a population inversion will occur between the E_3 and E_2 energy states, and a photon entering the population will be amplified coherently.

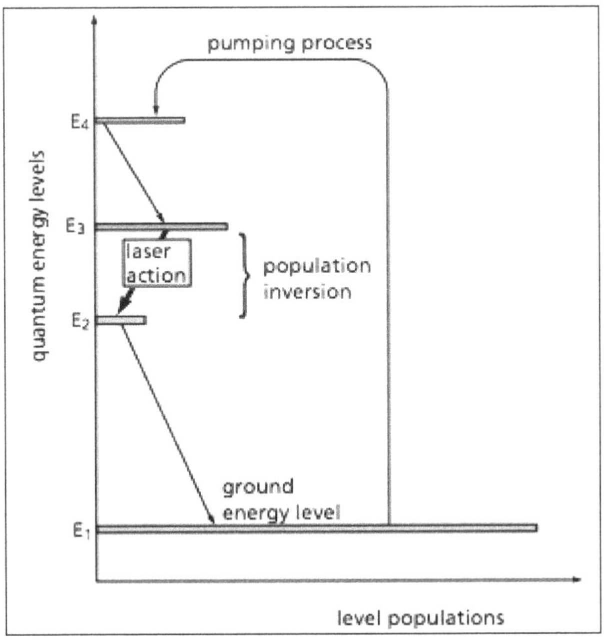

A four-level laser pumping system.

The Resonator

Although with a population inversion we have the ability to amplify a signal via stimulated emission, the overall single-pass gain is quite small, and most of the excited atoms in the population emit spontaneously and do not contribute to the overall output. To turn this system into a laser, we need a positive feedback mechanism that will cause the majority of the atoms in the population to contribute to the coherent output. This is the resonator, a system of mirrors that reflects undesirable (off-axis) photons out of the system and reflects the desirable (on-axis) photons back into the excited population where they can continue to be amplified.

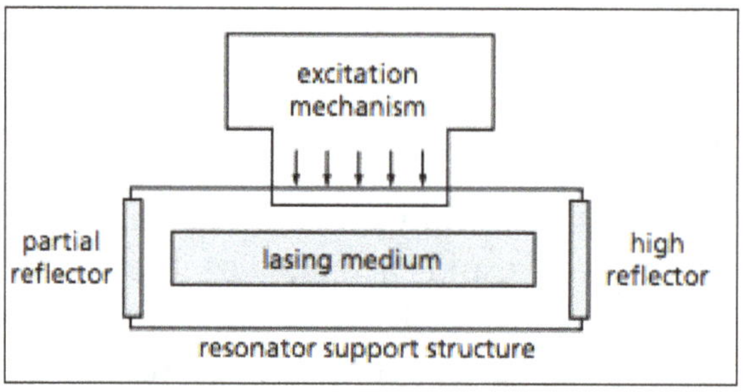

Schematic diagram of a basic laser.

Now consider the laser system shown in figure. The lasing medium is pumped continuously to create a population inversion at the lasing wavelength. As the excited atoms start to decay, they emit photons spontaneously in all directions. Some of the photons travel along the axis of the lasing medium, but most of the photons are directed out the sides. The photons traveling along the axis have an opportunity to stimulate atoms they encounter to emit photons, but the ones radiating out the sides do not. Furthermore, the photons traveling parallel to the axis will be reflected back into the lasing medium and given the opportunity to stimulate more excited atoms. As the on-axis photons are reflected back and forth interacting with more and more atoms, spontaneous emission decreases, stimulated emission along the axis predominates, and we have a laser.

Finally, to get the light out of the system, one of the mirrors is has a partially transmitting coating that couples out a small percentage of the circulating photons. The amount of coupling depends on the characteristics of the laser system and varies from a fraction of a percent for helium neon lasers to 50 percent or more for high-power lasers.

Practical Optical Coatings

In the design of a real-world laser, the optical resonator is often the most critical component, and, particularly for low-gain lasers, the most critical components of the resonator are the mirrors themselves. The difference between a perfect mirror coating (the optimum transmission and reflection with no scatter or absorption losses) and a real-world coating, capable of being produced in volume, can mean a 50-percent (or greater) drop in output power from the theoretical maximum. Consider the

543-nm green helium neon laser line. It was first observed in the laboratory in 1970, but, owing to its extremely low gain, the mirror fabrication and coating technology of the day was incapable of producing a sufficiently loss-free mirror that was also durable. Not until the late 1990s had the mirror coating technology improved sufficiently that these lasers could be offered commercially in large volumes.

The critical factors for a mirror, other than transmission and reflection, are scatter, absorption, stress, surface figure, and damage resistance. Coatings with low damage thresholds can degrade over time and cause output power to drop significantly. Coatings with too much mechanical stress not only can cause significant power loss, but can also induce stress birefringence, which can result in altered polarization and phase relationships. The optical designer must take great care when selecting the materials for the coating layers and the substrate to ensure that the mechanical, optical, and environmental characteristics are suitable for the application.

The equipment used for both substrate polishing and optical coating is a critical factor in the end result. Coating scatter is a major contributor to power loss. Scatter arises primarily from imperfections and inclusions in the coating, but also from minute imperfections in the substrate. Over the last few years, the availability of "super-polished" mirror substrates has led to significant gains in laser performance. Likewise, ion-beam sputtering and next-generation ion-assisted ion deposition has increased the packing density of laser coatings, thereby reducing absorption, increasing damage thresholds, and enabling the use of new and exotic coating materials.

Propagation Characteristics of Laser Beams

Beam Waist and Divergence

Diffraction causes light waves to spread transversely as they propagate, and it is therefore impossible to have a perfectly collimated beam. The spreading of a laser beam is in accord with the predictions of diffraction theory. Under ordinary circumstances, the beam spreading can be so small it can go unnoticed. The following formulas accurately describe beam spreading, making it easy to see the capabilities and limitations of laser beams. Even if a Gaussian TEM_{00} laser-beam wavefront were made perfectly flat at some plane, with all rays there moving in precisely parallel directions, it would acquire curvature and begin spreading in accordance with,

$$R(z) = z\left[1 + \left(\frac{\pi w_0^{\;2}}{\lambda z}\right)^2\right]$$

and

$$w(z) = w_0\left[1 + \left(\frac{\lambda z}{\pi w_0^{\;2}}\right)^2\right]^{1/2}$$

Where, z is the distance propagated from the plane where the wavefront is flat, λ is the wavelength of light, w_0 is the radius of the $1/e^2$ irradiance contour at the plane where the wavefront is flat, w(z) is the radius of the $1/e^2$ contour after the wave has propagated a distance z, and R(z) is the wavefront radius of curvature after propagating a distance z. R(z) is infinite at z = 0, passes through a minimum at some finite z, and rises again toward infinity as z is further increased, asymptotically approaching the value of z itself. The plane z = 0 marks the location of a beam waist, or a place where the wave front is flat, and w_0 is called the beam waist radius. The irradiance distribution of the Gaussian TEM_{00} beam, namely,

$$I(r) = I_0 e^{-2r^2/w^2} = \frac{2P}{\pi w^2} e^{-2r^2/w^2},$$

Where, $w = w(z)$ and P is the total power in the beam, is the same at all cross sections of the beam. The invariance of the form of the distribution is a special consequence of the presumed Gaussian distribution at z = 0. Simultaneously, as R(z) asymptotically approaches z for large z, $w(z)$ asymptotically approaches the value,

$$w(z) = \frac{\lambda z}{\pi w_0}$$

where, z is presumed to be much larger than $\pi w_0^{\;2}/\lambda$ so that the $1/e^2$ irradiance contours asymptotically approach a cone of angular radius,

$$\theta = \frac{w(z)}{z} = \frac{\lambda}{\pi w_0}$$

This value is the far-field angular radius (half-angle divergence) of the Gaussian TEM00 beam. The vertex of the cone lies at the center of the waist. It is important to note that, for a given value of λ, variations of beam diameter and divergence with distance z are functions of a single parameter, w_0, the beam waist radius.

Near-field vs. Far-field Divergence

Unlike conventional light beams, Gaussian beams do not diverge linearly. Near the laser, the divergence angle is extremely small; far from the laser, the divergence angle approaches the asymptotic limit described in above equation. The Raleigh range (z_R), defined as the distance over which the beam radius spreads by a factor of √2, is given by,

$$z_R = \frac{\pi w_0^2}{\lambda}$$

The Raleigh range is the dividing line between near-field divergence and mid-range divergence. Far-field divergence (the number quoted in laser specifications) must be measured at a point $>z_R$ (usually $10z_R$ will suffice). This is a very important distinction because calculations for spot size and other parameters in an optical train will be inaccurate if near- or mid-field divergence values are used. For a tightly focused beam, the distance from the waist (the focal point) to the far field can be a few millimeters or less. For beams coming directly from the laser, the far-field distance can be measured in meters.

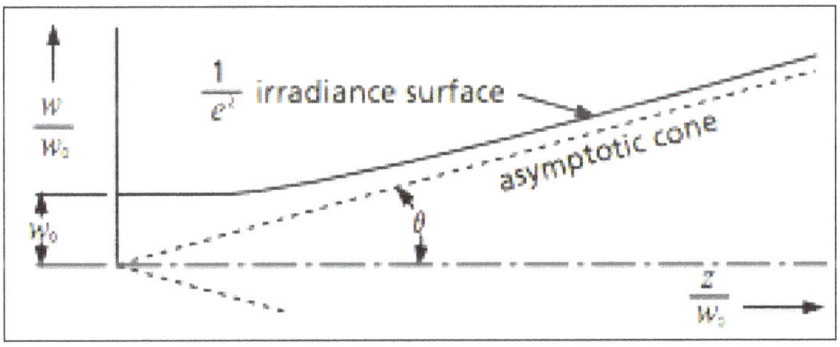

Growth in beam diameter as a function of distance from the beam waist.

Locating the Beam Waist

For a Gaussian laser beam, the location (and radius) of the beam waist is determined uniquely by the radius of curvature and optical spacing of the laser cavity mirrors because, at the reflecting surfaces of the cavity mirrors, the radius of curvature of the propagating beam is exactly the same as that of the mirrors. Consequently, for the flat/curved cavity shown in figure, the beam waist is located at the surface of the flat mirror. For a symmetric cavity (b), the beam waist is halfway between the mirrors; for non-symmetric cavities (c and d), the beam waist is located by using the equation,

$$z_1 = \frac{L(R_2 - L)}{R_1 + R_2 - 2L}$$

and

$$z_1 + z_2 = L$$

where, L is the effective mirror spacing, R_1 and R_2 are the radii of curvature of the cavity mirrors, and z_1 and z_2 are the distances from the beam waist of mirrors 1 and 2, respectively.

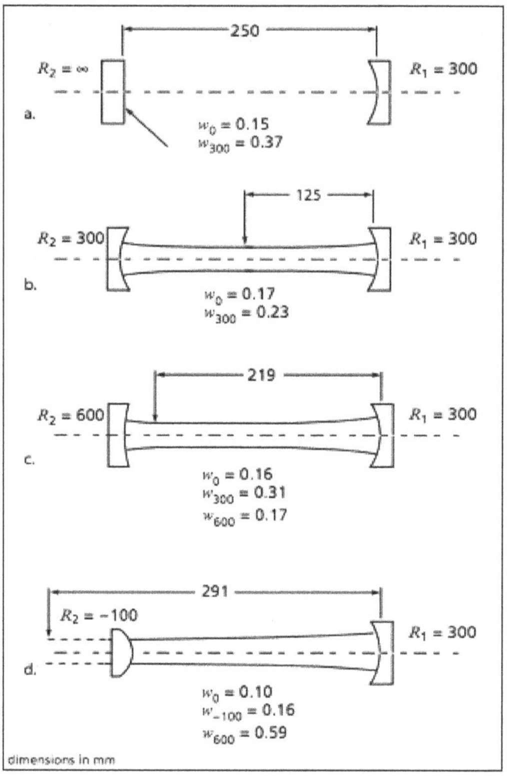

Location of beam waist for common cavity geometries.

In any case but that of a flat output mirror, the beam waist is refracted as it passes through the mirror substrate. If the output coupler's second surface is flat, the effective waist of the refracted beam is moved toward the output coupler and is reduced in diameter. However, by applying a spherical correction to the second surface of the output coupler, the location of the beam waist can be moved to the output coupler itself, increasing the beam waist diameter and reducing far-field divergence.

It is useful, particularly when designing laser cavities, to understand the effect that mirror spacing has on the beam radius, both at the waist and at the curved mirror. Figure plots equations $R(z) = z\left[1+\left(\frac{\pi w_0^2}{\lambda z}\right)^2\right]$ and $w(z) = w_0\left[1+\left(\frac{\lambda z}{\pi w_0^2}\right)^2\right]^{1/2}$ as a function of R/z (curved mirror radius divided by the mirror spacing). As the mirror spacing approaches the radius of curvature of the mirror (R/z = 1), the beam waist decreases dramatically, and the beam radius at the curved mirror becomes very large. On the other hand, as R/z becomes large, the beam radius at the waist and at the curved mirror are approximately the same.

Calculating a Correcting Surface

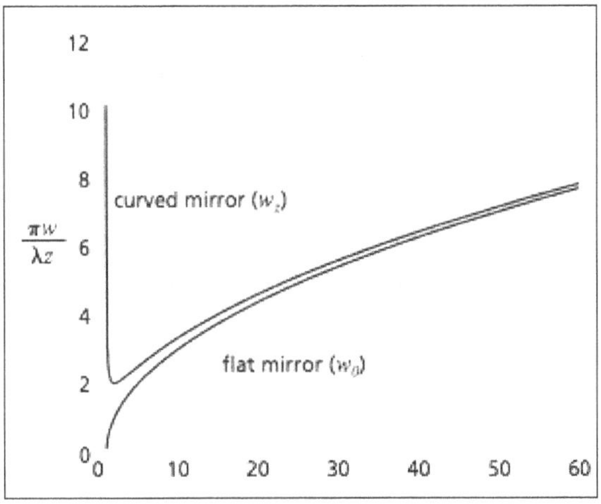

Beam waist and output diameter as a function of mirror radius and separation.

A laser beam is refracted as it passes through a curved output mirror. If the mirror has a flat second surface, the waist of the refracted beam moves closer to the mirror, and the divergence is increased. To counteract this, laser manufacturers often put a radius on the output coupler's second surface to collimate the beam by making a waist at the output coupler. This is illustrated by the case of a typical helium neon laser cavity consisting of a flat high reflector and an output mirror with a radius of curvature of 20 cm separated by 15 cm. If the laser is operating at 633 nm, the beam waist radius, beam radius at the output coupler, and beam half-angle divergence are:

$$w_0 = 0.13 \text{ mm}, \; w_{200} = 0.26 \text{ mm, and } \theta = 1.5 \text{ mrad},$$

However, with a flat second surface, the divergence nearly doubles to 2.8 mrad. Geometrical optics would give the focal length of the lens formed by the correcting output coupler as 15 cm; a rigorous calculation using Gaussian beam optics shows it should be 15.1 cm. Using the lens-makers formula,

$$\frac{1}{f} = (n-1)\left(\frac{1}{R_1} - \frac{1}{R_2}\right)$$

The appropriate sign convention and assuming that n = 1.5, we get a convex correcting curvature of approximately 5.5 cm. At this point, the beam waist has been transferred to the output coupler, with a radius of 0.26 mm, and the far-field half-angle divergence is reduced to 0.76 mrad, a factor of nearly 4.

Correcting surfaces are used primarily on output couplers whose radius of curvature is a meter or less. For longer radius output couplers, the refraction effects are less dramatic, and a correcting second surface radius is unnecessary.

Higher Order Gaussian Laser Beams

In the real world, the truly 100-percent, single transverse mode, Gaussian laser beam (also called a pure or fundamental mode beam) described by equations $R(z)=z\left[1+\left(\frac{\pi w_0^2}{\lambda z}\right)^2\right]$ and $w(z)=w_0\left[1+\left(\frac{\lambda z}{\pi w_0^2}\right)^2\right]^{1/2}$ is very hard to find. Low-power beams from helium neon lasers can be a close approximation, but the higher the power of the laser, and the more complex the excitation mechanism (e.g., transverse discharges, flash-lamp pumping), or the higher the order of the mode, the more the beam deviates from the ideal.

To address the issue of higher order Gaussian beams and mixed mode beams, a beam quality factor, M^2, has come into general use. A mixed mode is one where several modes are oscillating in the resonator at the same time. A common example is the mixture of the lowest order single transverse mode with the doughnut mode, before the intracavity mode limiting aperture is critically set to select just the fundamental mode. Because all beams have some wavefront defects, which implies they contain at least a small admixture of some higher order modes, a mixed mode beam is also called a "real" laser beam.

For a theoretical single transverse mode Gaussian beam, the value of the waist radius–divergence product is:

$$w_0\theta = \lambda / \pi$$

It is important to note that this product is an invariant for transmission of a beam through any normal, high-quality optical system (one that does not add aberrations to the beam wavefront). That is, if a lens focuses the single mode beam to a smaller waist radius, the convergence angle coming into the focus (and the divergence angle emerging from it) will be larger than that of the unfocused beam in the same ratio that the focal spot diameter is smaller: the product is invariant. For a real laser beam we have,

$$W_0\theta = M^2\lambda / \pi$$

where, w_0 and θ are the $1/e^2$ intensity waist radius and the far-field half-divergence angle of the real laser beam, respectively. Here we have introduced the convention that upper case symbols are used for the mixed mode and lower case symbols for the fundamental mode beam coming from the same resonator. The mixed-mode beam radius W is M times larger than the fundamental mode radius at all propagation distances. Thus the waist radius is that much larger, contributing the first factor of M in equation $w_0\theta = M^2\lambda / \pi$. The second factor of M comes from the half-angle divergence, which is also M times larger. The waist radius–divergence half-angle product for the mixed mode beam also is an invariant, but is M^2 larger. The fundamental mode beam has the smallest divergence allowed by diffraction for a beam of that waist radius. The factor

M^2 is called the "times-diffraction-limit" number or (inverse) beam quality; a diffraction-limited beam has an M^2 of unity.

For a typical helium neon laser operating in TEM00 mode, $M^2 < 1.05$. Ion lasers typically have an M^2 factor ranging from 1.1 to 1.7. For high-energy multimode lasers, the M^2 factor can be as high as 30 or 40. The M^2 factor describes the propagation characteristics (spreading rate) of the laser beam. It cannot be neglected in the design of an optical train to be used with the beam. Truncation (aperturing) by an optic, in general, increases the M^2 factor of the beam.

The propagation equations for the mixed-mode beam W(z) and R(z) are as follows:

$$W(z) = W_0\left[1+\left(\frac{zM^2\lambda}{\pi W_0^2}\right)^2\right]^{1/2} = W_0\left[1+\left(\frac{z}{Z_R}\right)^2\right]$$

$$R(z) = z\left[1+\left(\frac{\pi W_0^2}{zM^2\lambda}\right)^2\right] = z\left[1+\left(\frac{Z_R}{z}\right)^2\right].$$

The Rayleigh range remains the same for a mixed mode laser beam:

$$Z_R = \frac{\pi W_0^2}{M^2\lambda} = \frac{\pi w_0^2}{\lambda} = z_R.$$

Now consider the consequences in coupling a high M^2 beam into a fiber. Fiber coupling is a task controlled by the product of the focal diameter ($2W_f$) and the focal convergence angle (θ_f). In the tight focusing limit, the focal diameter is proportional to the focal length f of the lens, and is inversely proportional to the diameter of the beam at the lens (i.e., $2W_f \propto f/D_{\text{lens}}$).

The lens-to-focus distance is f, and, since $f \times \theta_f$ is the beam diameter at distance f in the far field of the focus, $D_{\text{lens}} \propto f\theta_f$. Combining these proportionalities yields,

$$W_f\theta_f = \text{constant}$$

For the fiber-coupling problem as stated above. The diameter divergence product for the mixed-mode beam is M^2 larger than the fundamental mode beam in accordance with equation $w_0\theta = \lambda/\pi$ and $W_0\theta = M^2\lambda/\pi$.

There is a threefold penalty associated with coupling a beam with a high M^2 into a fiber: 1) the focal length of the focusing lens must be a factor of $1/M^2$ shorter than that used with a fundamental-mode beam to obtain the same focal diameter at the fiber; 2) the numerical aperture (NA) of the focused beam will be higher than that of the fundamental beam (again by a factor of $1/M^2$) and may exceed the NA of the fiber; and 3) the depth of focus will be smaller by $1/M^2$ requiring a higher degree of precision and stability in the optical alignment.

Transverse Modes and Mode Control

The fundamental TEM_{00} mode is only one of many transverse modes that satisfies the condition that it be replicated each round-trip in the cavity. Figure shows examples of the primary lower-order Hermite-Gaussian (rectangular) modes. Note that the subscripts m and n in the mode designation TEM_{mn} are correlated to the number of nodes in the x and y directions. The propagation equation can also be written in cylindrical form in terms of radius (ρ) and angle (ϕ). The eigenmodes ($E_{\rho\phi}$) for this equation are a series of axially symmetric modes, which, for stable resonators, are closely approximated by Laguerre-Gaussian functions, denoted by $TEM_{\rho\phi}$. For the lowest-order mode, TEM_{00}, the Hermite-Gaussian and Laguerre-Gaussian functions are identical, but for higher-order modes, they differ significantly, as shown in figure.

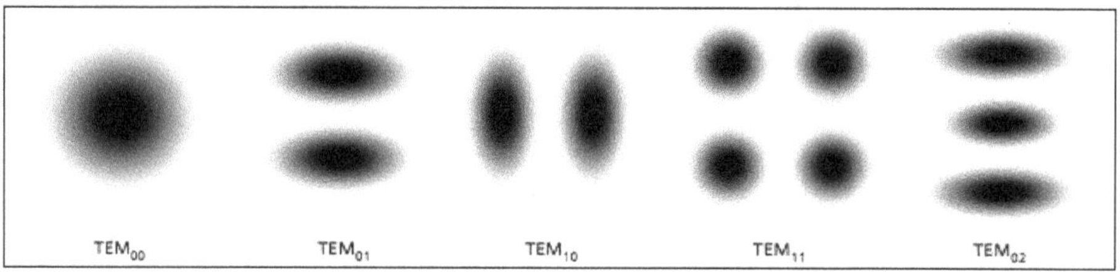

Low-order Hermite-Gaussian resonator modes.

The mode, TEM_{01}, also known as the "bagel" or "doughnut" mode, is considered to be a superposition of the HermiteGaussian TEM_{10} and TEM_{01} modes, locked in phase and space quadrature. In real-world lasers, the Hermite-Gaussian modes predominate since strain, slight misalignment, or contamination on the optics tends to drive the system toward rectangular coordinates. Nonetheless, the Laguerre-Gaussian TEM_{10} "target" or "bulls-eye" mode is clearly observed in well-aligned gas-ion and helium neon lasers with the appropriate limiting apertures.

Mode Control

The transverse modes for a given stable resonator have different beam diameters and divergences. The lower the order of the mode is, the smaller the beam diameter, the narrower the far-field divergence, and the lower the M^2 value. For example, the TEM_{01} doughnut mode is approximately 1.5 times the diameter of the fundamental TEM_{00} mode, and the Laguerre TEM_{10} target mode is twice the diameter of the TEM_{00} mode. The theoretical M^2 values for the TEM_{00}, TEM_{01}, and TEM_{10} modes are 1.0, 2.3, and 3.6, respectively. Because of its smooth intensity profile, low divergence, and ability to be focused to a diffraction-limited spot, it is usually desirable to operate in the lowest-order mode possible, TEM_{00}. Lasers, however, tend to operate at the highest-order mode possible, either in addition to, or instead of, TEM_{00} because the larger beam diameter may allow them to extract more energy from the lasing medium.

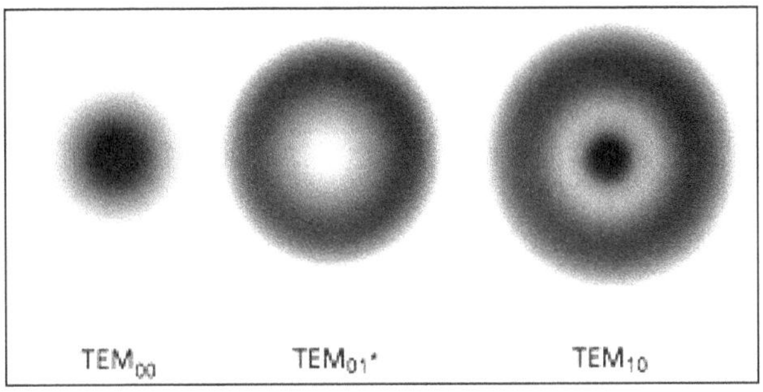

Low-order axisymmetric resonator modes.

The primary method for reducing the order of the lasing mode is to add sufficient loss to the higher-order modes so that they cannot oscillate without significantly increasing the losses at the desired lower-order mode. In most lasers this is accomplished by placing a fixed or variable aperture inside the laser cavity. Because of the significant differences in beam diameter, the aperture can cause significant diffraction losses for the higher-order modes without impacting the lower-order modes. As an example, consider the case of a typical argon-ion laser with a long-radius cavity and a variable mode-selecting aperture.

When the aperture is fully open, the laser oscillates in the axially symmetric TEM_{10} target mode. As the aperture is slowly reduced, the output changes smoothly to the TEM_{01} doughnut mode, and finally to the TEM_{00} fundamental mode. In many lasers, the limiting aperture is provided by the geometry of the laser itself. For example, by designing the cavity of a helium neon laser so that the diameter of the fundamental mode at the end of the laser bore is approximately 60 percent of the bore diameter, the laser will naturally operate in the TEM_{00} mode.

Single Axial Longitudinal Mode Operation

Theory of Longitudinal Modes

In a laser cavity, the requirement that the field exactly reproduce itself in relative amplitude and phase each round-trip means that the only allowable laser wavelengths or frequencies are given by,

$$\lambda = \frac{P}{N} \; or \; v = \frac{Nc}{P}$$

where λ is the laser wavelength, ν is the laser frequency, c is the speed of light in a vacuum, N is an integer whose value is determined by the lasing wavelength, and P is the effective perimeter optical path length of the beam as it makes one round-trip, taking into account the effects of the index of refraction. For a conventional two-mirror cavity in which the mirrors are separated by optical length L, these formulas revert to the familiar,

$$\lambda = \frac{2L}{N} or \; v = \frac{Nc}{2L}$$

These allowable frequencies are referred to as longitudinal modes. The frequency spacing between adjacent longitudinal modes is given by,

$$\Delta v = \frac{c}{P}.$$

As can be seen from above equation, the shorter the laser cavity is, the greater the mode spacing will be. By differentiating the expression for n with respect to P we arrive at,

$$\delta v = -\frac{Nc}{P^2}\delta P \; or \; \delta n = -\frac{Nc}{2L^2}\delta L.$$

Consequently, for a helium neon laser operating at 632.8 nm, with a cavity length of 25 cm, the mode spacing is approximately 600 MHz, and a 100-nm change in cavity length will cause a given longitudinal mode to shift by approximately 190 MHz.

The number of longitudinal laser modes that are present in a laser depends primarily on two factors: the length of the laser cavity and the width of the gain envelope of the lasing medium. For example, the gain of the red helium neon laser is centered at 632.8 nm and has a full width at half maximum (FWHM) of approximately 1.4 GHz, meaning that, with a 25-cm laser cavity, only two or three longitudinal modes can be present simultaneously, and a change in cavity length of less than one micron will cause a given mode to "sweep" completely through the gain. Doubling the cavity length doubles the number of oscillating longitudinal modes that can fit under the gain curve doubles.

The gain of a gas-ion laser (e.g., argon or krypton) is approximately five times broader than that of a helium neon laser, and the cavity spacing is typically much greater, allowing many more modes to oscillate simultaneously.

A mode oscillating at a frequency near the peak of the gain will extract more energy from the gain medium than one oscillating at the fringes. This has a significant impact on the performance of a laser system because, as vibration and temperature changes cause small changes in the cavity length, modes sweep back and forth through the gain. A laser operating with only two or three longitudinal modes can experience power fluctuations of 10% or more, whereas a laser with ten or more longitudinal modes will see mode-sweeping fluctuations of 2 percent or less.

Selecting a Single Longitudinal Mode

A laser that operates with a single longitudinal mode is called a single-frequency laser. There are two ways to force a conventional two-mirror laser to operate with a single longitudinal mode. The first is to design the laser with a short enough cavity that only a single mode can be sustained. For example, in the helium neon laser described above, a 10-cm cavity would allow only one mode to oscillate. This is not a practical approach for most gas lasers because, with the cavity short enough to suppress additional modes, there may be insufficient energy in the lasing medium to sustain any lasing action at all, and if there is lasing, the output will be very low.

The second method is to introduce a frequency-control element, typically a low-finesse Fabry-Perot etalon, into the laser cavity. The free spectral range of the etalon should be several times the width of the gain curve, and the reflectivity of the surfaces should be sufficient to provide 10 percent or greater loss at frequencies half a longitudinal mode spacing away from the etalon peak. The etalon is mounted at a slight angle to the optical axis of the laser to prevent parasitic oscillations between the etalon surfaces and the laser cavity.

Once the mode is selected, the challenge is to optimize and maintain its output power. Since the laser mode moves if the cavity length changes slightly, and the etalon pass band shifts if the etalon spacing varies slightly, it important that both be stabilized. Various mechanisms are used. Etalons can be passively stabilized by using zero-expansion spacers and thermally stabilized designs, or they can be thermally stabilized by placing the etalon in a precisely controlled oven. Likewise, the overall laser cavity can be passively stabilized, or, alternatively, the laser cavity can be actively stabilized by providing a servomechanism to control cavity length, as discussed in Frequency Stabilization.

The Ring Laser: The discussions above are limited to twomirror standing-wave cavities. Some lasers operate naturally in a single longitudinal mode. For example, a ring laser cavity, (used in many dye and Ti:Sapphire lasers as well as in gyroscopic lasers) that has been constrained to oscillate in only one direction produces a traveling wave without the fixed nodes of the standing-wave laser. The traveling wave sweeps through the laser gain, utilizing all of the available energy and preventing the buildup of adjacent modes. Other lasers are "homogeneously broadened" allowing virtually instantaneous transfer of energy from one portion of the gain curve to another.

Frequency Stabilization

The frequency output of a single-longitudinal-mode laser is stabilized by precisely controlling the laser cavity length. This can be accomplished passively by building an athermalized resonator structure and carefully controlling the laser environment to eliminate expansion, contraction, and vibration, or actively by using a mechanism to determine the frequency (either relatively or absolutely) and quickly adjusting the laser cavity length to maintain the frequency within the desired parameters.

A typical stabilization scheme is shown in figure. A portion of the laser output beam is directed into a low-finesse Fabry-Perot etalon and tuned to the side of the transmission band. The throughput is compared to a reference beam, as shown in the figure. If the laser frequency increases, the ratio of attenuated power to reference power increases. If the laser frequency decreases, the ratio decreases. In other words, the etalon is used to create a frequency discriminant that converts changes in frequency to changes in power. By "locking" the discriminant ratio at a specific value (e.g., 50 percent) and providing negative feedback to the device used to control cavity length, output frequency can be controlled. If the frequency increases from the preset value, the length of the laser cavity is increased to drive the frequency back to the set point. If the frequency decreases, the cavity length is decreased. The response time of the control electronics is determined by the characteristics of the laser system being stabilized.

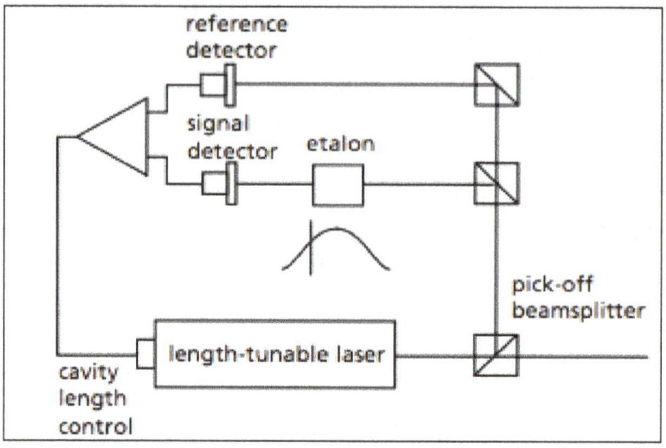

Laser frequency stabilization scheme.

Other techniques can be used to provide a discriminant. One common method used to provide an ultrastable, long-term reference is to replace the etalon with an absorption cell and stabilize the system to the saturated center of an appropriate transition. Another method, shown in figure is used with commercial helium neon lasers. It takes advantage of the fact that, for an internal mirror tube, the adjacent modes are orthogonally polarized. The cavity length is designed so that two modes can oscillate under the gain curve. The two modes are separated outside the laser by a polarization-sensitive beamsplitter. Stabilizing the relative amplitude of the two beams stabilizes the frequency of both beams.

The cavity length changes needed to stabilize the laser cavity are very small. In principle, the maximum adjustment needed is that required to sweep the frequency through one free spectral range of the laser cavity (the cavity mode spacing). For the helium neon laser cavity described earlier, the required change is only 320 nm, well within the capability of piezoelectric actuators. Commercially available systems can stabilize frequency output to 1 MHz or less. Laboratory systems that stabilize the frequency to a few kilohertz have been developed.

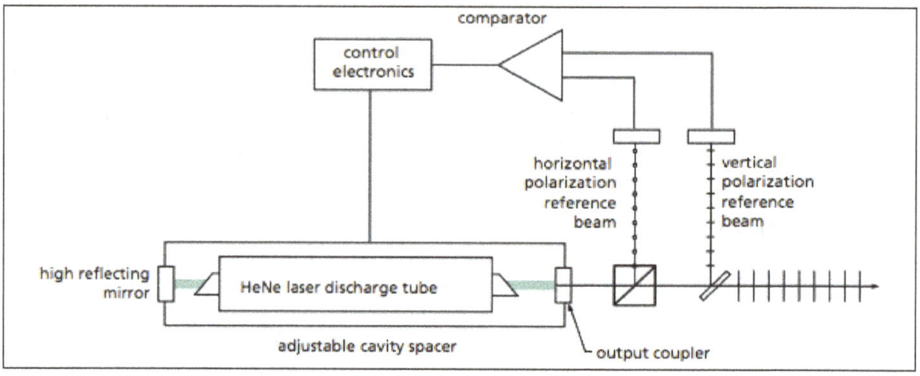

Frequency stabilization for a helium neon laser.

Frequency and Amplitude Fluctuations

The output of a freely oscillating laser will fluctuate in both amplitude and frequency. Fluctuations of less than 0.1 Hz are commonly referred to as "drift"; faster fluctuations are termed "noise" or, when talking about sudden frequency shifts, "jitter."

The major sources of noise in a laser are fluctuations in the pumping source and changes in length or alignment caused by vibration, stress, and changes in temperature. For example, unfiltered line ripple can cause output fluctuations of 5 to 10 percent or more. Likewise, a 10-µrad change in alignment can cause a 10-percent variation in output power, and, depending upon the laser, a 1-µm change in length can cause amplitude fluctuations of up to 50 percent (or more) and frequency fluctuations of several gigahertz.

High-frequency noise (>1 MHz) is caused primarily by "mode beating." Transverse Laguerre-Gaussian modes of adjacent order are separated by a calculable fraction of the longitudinal mode spacing, typically ~17 MHz in a 1-m resonator with long radius mirrors. If multiple transverse modes oscillate simultaneously, heterodyne interference effects, or "beats," will be observed at the difference frequencies. Likewise, mode beating can occur between longitudinal modes at frequencies of,

$$\Delta v_{longitudinal} = \frac{c}{2L} = \frac{c}{2P}.$$

Mode beating can cause peak-to-peak power fluctuations of several percent. The only way to eliminate this noise component is to limit the laser output to a single transverse and single longitudinal mode.

Finally, when all other sources of noise have been eliminated, we are left with quantum noise, the noise generated by the spontaneous emission of photons from the upper laser level in the lasing medium. In most applications, this is inconsequential.

Methods for Suppressing Amplitude Noise and Drift

Two primary methods are used to stabilize amplitude fluctuations in commercial lasers: automatic current control (ACC), also known as current regulation, and automatic power control (APC), also known as light regulation. In ACC, the current driving the pumping process passes through a stable sensing resistor, as shown in figure and the voltage across this resistor is monitored. If the current through the resistor increases, the voltage drop across the resistor increases proportionately. Sensing circuitry compares this voltage to a reference and generates an error signal that causes the power supply to reduce the output current appropriately. If the current decreases, the inverse process occurs. ACC is an effective way to reduce noise generated by the power supply, including line ripple and fluctuations.

With APC, instead of monitoring the voltage across a sensing resistor, a small portion of the output power in the beam is diverted to a photo detector, as shown in figure and the voltage generated by the detector circuitry is compared to a reference. As output power fluctuates, the sensing circuitry generates an error signal that is used to make the appropriate corrections to maintain constant output.

Automatic current control effectively reduces amplitude fluctuations caused by the driving electronics, but it has no effect on amplitude fluctuations caused by vibration or misalignment. Automatic power control can effectively reduce power fluctuations from all sources. Neither of these control mechanisms has a large impact on frequency stability.

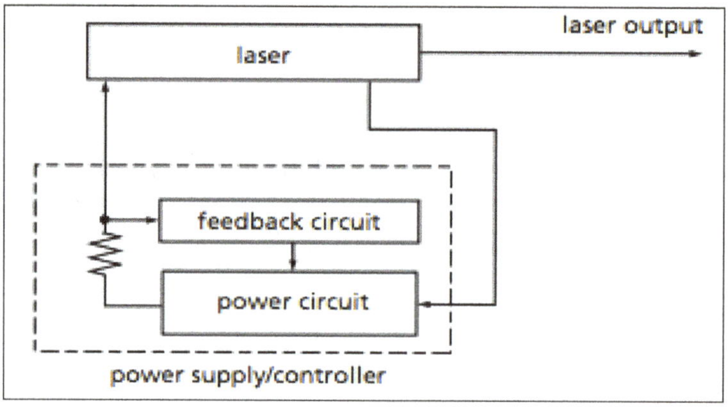

Automatic current control schematic.

Not all continuous-wave lasers are amenable to APC as described above. For the technique to be effective, there must be a monotonic relationship between output power and a controllable parameter (typically current or voltage). For example, throughout the typical operating range of a gas-ion laser, an increase in current will increase the output power and vice versa. This is not the case for some lasers. The output of a helium neon laser is very insensitive to discharge current, and an increase in current may increase or decrease laser output. In a helium cadmium laser, where electrophoresis determines the density and uniformity of cadmium ions throughout the discharge, a slight change in discharge current in either direction can effectively kill lasing action.

If traditional means of APC are not suitable, the same result can be obtained by placing an acousto-optic modulator inside the laser cavity and using the error signal to control the amount of circulating power ejected from the cavity.

One consideration that is often overlooked in an APC system is the geometry of the light pickoff mechanism itself. One's first instinct is to insert the pickoff optic into the main beam at a 45-degree angle, so that the reference beam exits at a 90-degree angle. However, as shown in figure for uncoated glass, there is almost a 10-percent difference in reflectivity for *s* and *p* polarization.

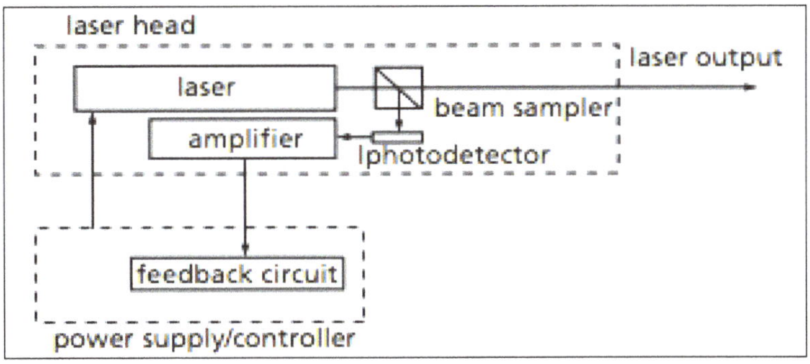

Automatic power control schematic.

In a randomly polarized laser, the ratio of the s and p components is not necessarily stable, and using a 90-degree reference beam can actually increase amplitude fluctuations. This is of much less concern in a laser with a high degree of linear polarization (e.g., 500:1 or better), but even then there is a slight presence of the orthogonal polarization. Good practice dictates that the pickoff element be inserted at an angle of 25 degrees or less.

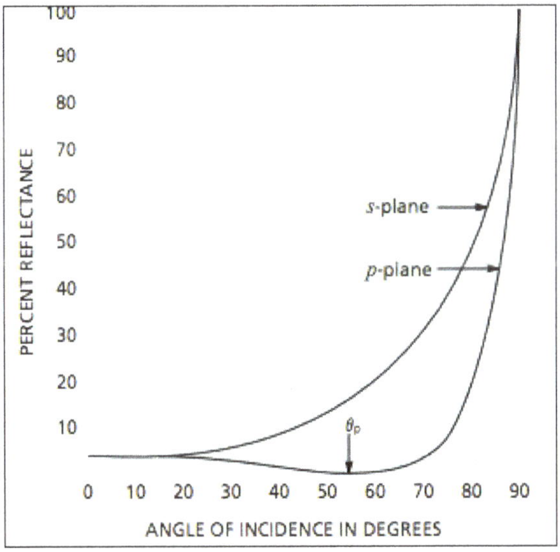

Reflectivity of a glass surface vs. incidence angle for *s* and *p* polarization.

Tunable Operation

Many lasers can operate at more than one wavelength. Argon and krypton lasers can operate at discrete wavelengths ranging from the ultraviolet to the near infrared. Dye lasers can be continuously tuned over a spectrum of wavelengths determined by the fluorescence bandwidths of the specific dyes (typically about 150 nm). Alexandrite and titanium sapphire lasers can be tuned continuously over specific spectral regions.

To create a tunable laser, the cavity coatings must be sufficiently broadband to accommodate the entire tuning range, and a variable-wavelength tuning element must be introduced into the cavity, either between the cavity optics or replacing the high-reflecting optic, to introduce loss at undesired wavelengths.

Three tuning mechanisms are in general use: Littrow prisms, diffraction gratings, and birefringent filters. Littrow prisms and their close relative, the full-dispersing prism, are used extensively with gas lasers that operate at discrete wavelengths. In its simplest form, the Littrow prism is a 30-60-90-degree prism with the surface opposite the 60-degree angle coated with a broadband high-reflecting coating. The prism is oriented so that the desired wavelength is reflected back along the optical axis, and the other wavelengths are dispersed off axis. By rotating the prism the retroreflected wavelength can be changed. In laser applications, the prism replaces the high-reflecting mirror, and the prism's angles are altered (typically to 34, 56, and 90 degrees) to minimize intracavity losses by having the beam enter the prism exactly at Brewster's angle. For higher-power lasers which require greater dispersion to separate closely spaced lines, the Littrow prism can be replaced by a full-dispersing prism coupled with a high reflecting mirror.

Gratings are used for laser systems that require a higher degree of dispersion than that of a full-dispersing prism.

Birefringent filters have come into general use for continuously tunable dye and Ti:Sapphire lasers, since they introduce significantly lower loss than do gratings. The filter is made from a thin, crystalline-quartz plate with its fast axis oriented in the plane of the plate. The filter, placed at Brewster's angle in the laser beam, acts like a weak etalon with a free spectral range wider than the gain curve of the lasing medium. Rotating the filter around the normal to its face shifts the transmission bands, tuning the laser. Since there are no coatings and the filter is at Brewster's angle (thereby polarizing the laser), there are no inherent cavity reflection losses at the peak of the transmission band. A single filter does not have as significant a line-narrowing effect as does a grating, but this can be overcome by stacking multiple filter plates together, with each successive plate having a smaller free spectral range.

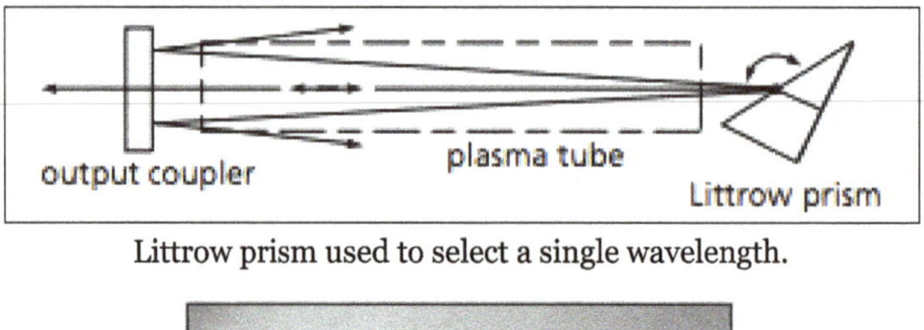

Littrow prism used to select a single wavelength.

Prism-tunable-ion laser.

Laser Construction

A laser is constructed from three principal parts:

- An energy source (usually referred to as the pump or pump source),
- A gain medium or laser medium, and
- Two or more mirrors that form an optical resonator.

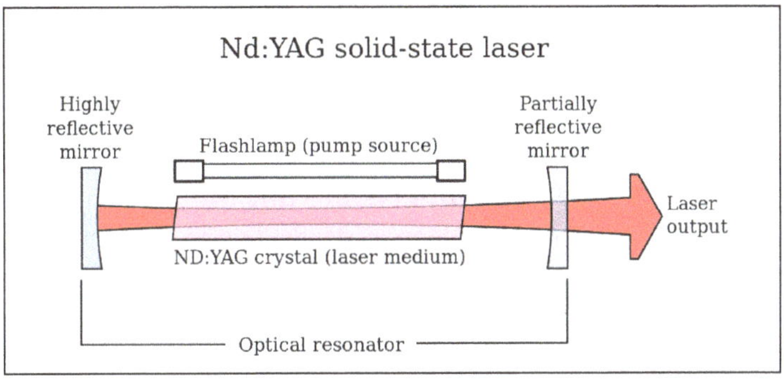

Schematic diagram of a typical laser, showing the three major.

Pump Source

The *pump source* is the part that provides energy to the laser system. Examples of pump sources include electrical discharges, flashlamps, arc lamps, light from another laser, chemical reactions and even explosive devices. The type of pump source used principally depends on the *gain medium*, and this also determines how the energy is transmitted to the medium. A helium–neon (HeNe) laser uses an electrical discharge in the helium-neon gas mixture, a Nd:YAG laser uses either light focused from a xenon flash lamp or diode lasers, and excimer lasers use a chemical reaction.

Gain Medium/Laser Medium

The gain medium is the major determining factor of the wavelength of operation, and other properties, of the laser. Gain media in different materials have linear spectra or wide spectra. Gain media with wide spectra allow tuning of the laser frequency. There are hundreds if not thousands of different gain media in which laser operation has been achieved. The gain medium is excited by the pump source to produce a population inversion, and it is in the gain medium where spontaneous and stimulated emission of photons takes place, leading to the phenomenon of optical gain, or amplification.

Examples of different gain media include:

- Liquids, such as dye lasers. These are usually organic chemical solvents, such as methanol, ethanol or ethylene glycol, to which are added chemical dyes such as coumarin, rhodamine, and fluorescein. The exact chemical configuration of the dye molecules determines the operation wavelength of the dye laser.

- Gases, such as carbon dioxide, argon, krypton and mixtures such as helium–neon. These lasers are often pumped by electrical discharge.
- Solids, such as crystals and glasses. The solid *host* materials are usually doped with an impurity such as chromium, neodymium, erbium or titanium ions. Typical hosts include YAG (yttrium aluminium garnet), YLF (yttrium lithium fluoride), sapphire (aluminium oxide) and various glasses. Examples of solid-state laser media include Nd:YAG, Ti:sapphire, Cr:sapphire (usually known as ruby), Cr:LiSAF (chromium-doped lithium strontium aluminium fluoride), Er:YLF, Nd:glass, and Er:glass. Solid-state lasers are usually pumped by flashlamps or light from another laser.
- Semiconductors, a type of solid, crystal with uniform dopant distribution or material with differing dopant levels in which the movement of electrons can cause laser action. Semiconductor lasers are typically very small, and can be pumped with a simple electric current, enabling them to be used in consumer devices such as compact disc players.

Optical Resonator

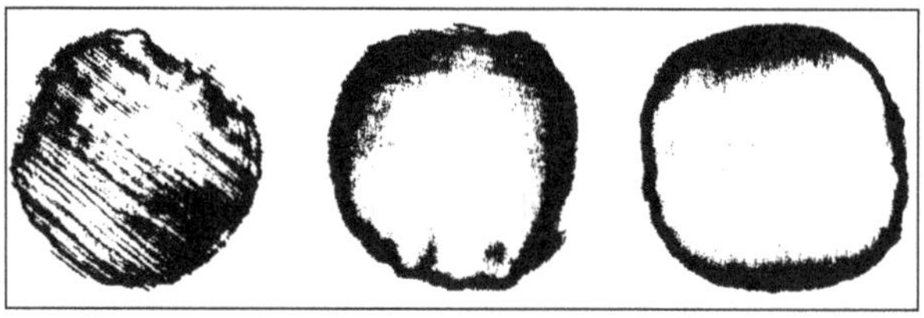

The Gaussian beam photographic paper burn comparison of a carbon dioxide transversely-excited atmospheric-pressure laser, obtained during the optimization process by adjusting the alignment mirrors.

The *optical resonator*, or *optical cavity*, in its simplest form is two parallel mirrors placed around the gain medium, which provide feedback of the light. The mirrors are given optical coatings which determine their reflective properties. Typically, one will be a high reflector, and the other will be a partial reflector. The latter is called the output coupler, because it allows some of the light to leave the cavity to produce the laser's output beam.

Light from the medium, produced by spontaneous emission, is reflected by the mirrors back into the medium, where it may be amplified by stimulated emission. The light may reflect from the mirrors and thus pass through the gain medium many hundreds of times before exiting the cavity. In more complex lasers, configurations with four or more mirrors forming the cavity are used. The design and alignment of the mirrors with respect to the medium is crucial for determining the exact operating wavelength and other attributes of the laser system.

Other optical devices, such as spinning mirrors, modulators, filters, and absorbers, may be placed within the optical resonator to produce a variety of effects on the laser output, such as altering the wavelength of operation or the production of pulses of laser light.

Some lasers do not use an optical cavity, but instead rely on very high optical gain to produce significant amplified spontaneous emission (ASE) without needing feedback of the light back into the gain medium. Such lasers are said to be superluminescent, and emit light with low coherence but high bandwidth. Since they do not use optical feedback, these devices are often not categorized as lasers.

Types of Lasers

Lasers can be broadly classified into four categories: gas discharge lasers, semiconductor diode lasers, optically pumped lasers, and "other," a category which includes chemical lasers, gas-dynamics lasers, x-ray lasers, combustion lasers, and others developed primarily for military applications. These lasers are not discussed further here.

Gas-discharge Lasers

In principle, gas-discharge lasers are inherently simple—fill a container with gas, put some mirrors around it, and strike a discharge. In practice, they are much more complex because the gas mix, discharge parameters, and container configuration must be specifically and carefully designed to create the proper conditions for a population inversion. Furthermore, careful consideration must be given to how the discharge will react with its container and with the laser optics. Finally, since the temperature of the gas can affect the discharge conditions, questions of cooling must be addressed.

Figure below shows a cutaway of a helium neon laser, one of the simplest gas-discharge lasers. An electrical discharge is struck between the anode and cathode. The laser bore confines the discharge, creating the current densities needed to create the inversion. In this example, the laser mirrors are mounted to the ends of the tube and are effectively part of the gas container. In other cases the mirrors are external to the container, and light enters and exits the chamber through Brewster's windows or extremely low-loss antireflection-coated normal windows. Because most gas-discharge lasers are operated at extremely low pressures, a getter is needed to remove the impurities generated by outgassing in the walls of the container or by erosion of the electrodes and bore caused by the discharge. The Brewster's window is used to linearly polarize the output of the laser.

The most common types of gas-discharge lasers are helium neon lasers, helium cadmium lasers (a metal-vapor laser), noble-gas ion lasers (argon, krypton), carbon-dioxide lasers, and the excimer-laser family.

Helium Neon Lasers

The helium neon (HeNe) laser, shown in figure, the second laser to be discovered, was the first to be used in volume applications. Today, millions of these lasers are in the field, and only semiconductor diode lasers are sold in greater quantity.

The HeNe laser operates in a high-voltage (kV), low-current (mA) glow discharge. Its most familiar output wavelength is 633 nm (red), but HeNe lasers are also available with output at 543 nm (green), 594 nm (yellow), 612 nm (orange), and 1523 nm (near infrared). Output power is low, ranging from a few tenths to tens of milliwatts, depending on the wavelength and size of the laser tube. Helium is the major constituent (85 percent) of the gas mixture, but it is the neon component that is the actual lasing medium. The glow discharge pumps the helium atoms to an excited state that closely matches the upper energy levels of the neon atoms.

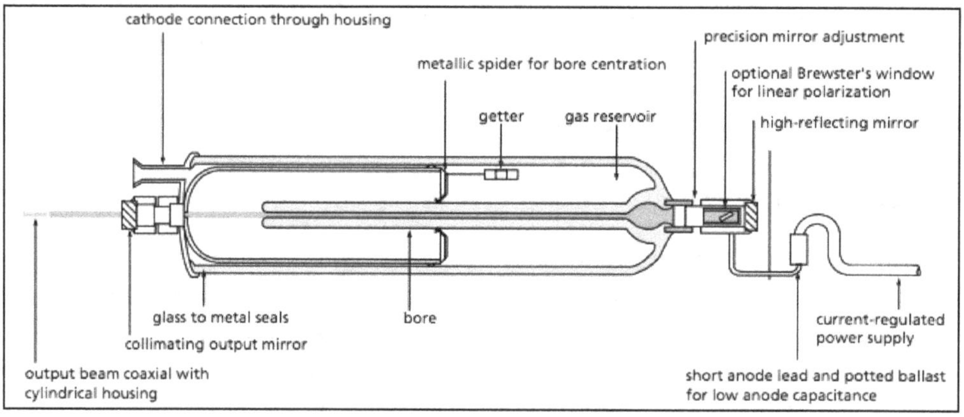

Typical HeNe laser construction.

This energy is then transferred to the neon atoms via collisions of the second kind (i.e., exciting the neon to a higher energy level as opposed to transferring the energy as kinetic motion). One characteristic of the glow discharge is its negative impedance (i.e., increasing the voltage decreases the current); consequently, to function with a standard current-regulated power supply, a ballast resistor must be used in series with the laser to make the overall impedance positive. The popularity (and longevity) of the HeNe laser is based on five factors: they are (relative to other lasers) small and compact; they have the best inherent beam quality of any laser, producing a virtually pure single transverse mode beam ($M^2 < 1.05$); they are extremely long lived, with many examples of an operating life of 50,000 hours or more; they generate relatively little heat and are convection cooled easily in OEM packages; and they have a relatively low acquisition and operating cost.

Helium Cadmium Lasers

Helium cadmium (HeCd) lasers are, in many respects, similar to the HeNe laser with the exception that cadmium metal, the lasing medium, is solid at room temperature.

The HeCd laser is a relatively economical, cw source for violet (442 nm) and ultraviolet (325 nm) output. Because of its excellent wavelength match to photopolymer and film sensitivity ranges, it is used extensively for three-dimensional stereolithography and holographic applications. As mentioned above, cadmium, a metal, is solid at room temperature. For lasing to occur, the metal must be evaporated from a reservoir, as shown in figure and then the vapor must be distributed uniformly down the laser bore. This is accomplished through a process called electrophoresis. Because cadmium will plate out on a cool surface, extreme care must be taken in the design of the laser to contain the cadmium and to protect the optics and windows from contamination, since even a slight film will introduce sufficient losses to stop lasing. The end of life usually occurs when cadmium is depleted from its reservoir.

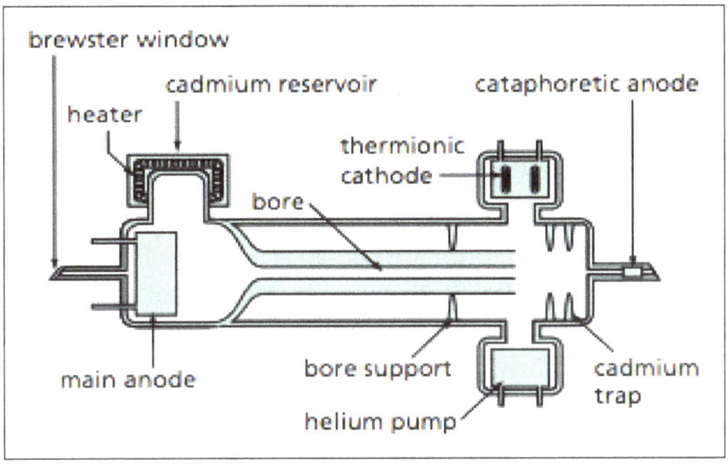

Construction of a HeCd laser.

Noble-gas Ion Lasers

The noble-gas ion lasers (argon-ion and krypton-ion), have been the mainstay of applications requiring high cw power in the visible, ultraviolet, and near-infrared spectral regions. High-power water-cooled systems can be found in research laboratories around the world; lower-power air-cooled systems are used in a wide variety of OEM applications. Argon-ion lasers are available with output up to 7 W in the ultraviolet (333–354 nm) and 25 W or more in the visible regions (454–515 nm), with primary output at 488 nm (blue) and 514 nm (green). Krypton-ion lasers have their primary output at 568 nm (yellow), 647 nm (red), and 752 nm (near infrared). Mixed-gas lasers combine both argon and krypton to produce lasers with a wider spectral coverage.

Unlike the HeNe laser, ion lasers operate with a high-intensity low-pressure arc discharge (low voltage, high current). A 20-W visible laser will require 10 kW or more power input, virtually all of which is deposited in the laser head as heat which must be removed from the system by some cooling mechanism. Furthermore, the current densities in the bore, which can be as high as $10^5 A/cm^2$, place large stresses on the bore materials.

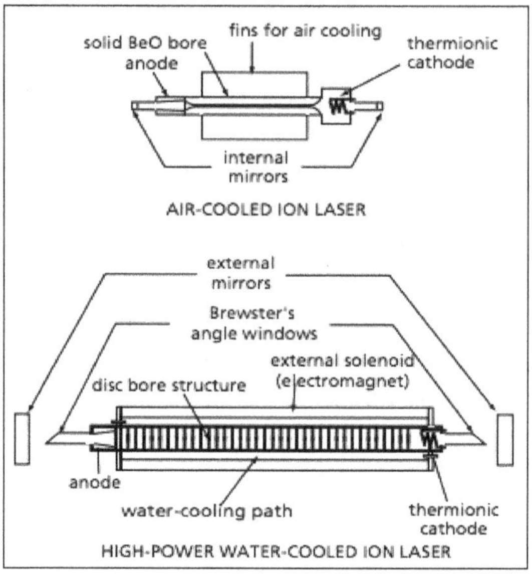

Air-cooled and water-cooled ion lasers.

Ion lasers can be broken into two groups: high-power (1–20=W) water-cooled lasers and low-power air-cooled lasers. Both are shown schematically in figure. The main features of both lasers are the same. Both use a coiled, directly-heated dispenser cathode to supply the current; both have a gas return path that counteracts gas pumping (non-uniform gas pressure throughout the length of the tube caused by the charged particles moving toward the electrodes).

The bore of an air-cooled system is always made of beryllium oxide (BeO), a ceramic known for its ability to conduct heat. A fin structure is attached to the outside of the ceramic bore, and a blower removes the generated heat, typically less than 1 kW. Water-cooled systems are available with either BeO bores or a construction wherein tungsten discs are attached to a thin-walled ceramic tube surrounded by a water jacket. The heat from the discs is conducted through the walls of the tube to the surrounding water. The entire bore structure is surrounded by a solenoid electromagnet, which compresses the discharge to increase current density and minimize bore erosion. The main life-limiting factors in ion lasers are cathode depletion and gas consumption. The intense discharge drives atoms into the walls of the discharge tube where they are lost to the discharge. Over time the tube pressure will decrease, causing the discharge to become unstable. This is particularly a problem with krypton-ion lasers. Water-cooled systems typically have some refill mechanism to keep the pressure constant. Air-cooled systems typically do not, limiting their practical operating life to approximately 5000 operating hours.

Carbon Dioxide Lasers

Because of their ability to produce very high power with relative efficiency, carbon dioxide (CO_2) lasers are used primarily for materials-processing applications. The standard output of these lasers is at 10.6 mm, and output power can range from less than 1 W to more than 10 kW.

Unlike atomic lasers, CO_2 lasers work with molecular transitions (vibrational and rotational states) which lie at low enough energy levels that they can be populated thermally, and an increase in the gas temperature, caused by the discharge, will cause a decrease in the inversion level, reducing output power. To counter this effect, high-power cw CO_2 lasers use flowing gas technology to remove hot gas from the discharge region and replace it with cooled (or cooler) gas. With pulsed CO_2 lasers that use transverse excitation, the problem is even more severe, because, until the heated gas between the electrodes is cooled, a new discharge pulse cannot form properly. A variety of types of CO_2 lasers are available. High-power pulsed and cw lasers typically use a transverse gas flow with fans which move the gas through a laminar-flow discharge region, into a cooling region, and back again. Low-power lasers most often use waveguide structures, coupled with radio-frequency excitation, to produce small, compact systems.

Excimer Lasers

The term excimer or "excited dimer" refers to a molecular complex of two atoms which is stable (bound) only in an electronically excited state. These lasers, which are available only as pulsed lasers, produce intense output in the ultraviolet and deep ultraviolet. The lasers in this family are XeFl (351 nm), XeCl (308 nm), KrF (248 nm), KrCl (222 nm), ArF (193 nm), and F_2 (157 nm). They are used extensively in photolithography, micromachining, and medical (refractive eye surgery) applications.

At first glance, the construction of an excimer laser is very similar to that of a transverse-flow, pulsed CO_2 laser. However, the major difference is that the gases in the system are extremely corrosive and great care must be taken in the selection and passivation of materials to minimize their corrosive effects. A system built for CO_2 would fail in minutes, if not seconds.

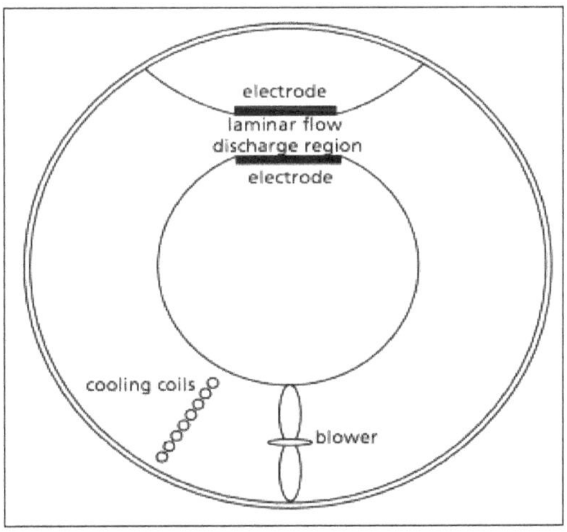

Schematics of transverse flow CO_2 laser system.

The principal advantage of an excimer laser is its very short wavelength. The excimer output beam can be focused to a spot diameter that is approximately 40 times smaller than the CO_2 laser beam with the same beam quality. Furthermore, whereas the long CO_2 wavelength removes material thermally via evaporation (boiling off material), the excimer lasers with wavelengths near 200 nm remove material via ablation (breaking molecules apart), without any thermal damage to the surrounding material.

Semiconductor Diode Lasers

The means of generating optical gain in a diode laser, the recombination of injected holes and electrons (and consequent emission of photons) in a forward-biased semiconductor pn junction, represents the direct conversion of electricity to light. This is a very efficient process, and practical diode laser devices reach a 50-percent electrical-to-optical power conversion rate, at least an order of magnitude larger than most other lasers. Over the past 20 years, the trend has been one of a gradual replacement of other laser types by diode laser based–solutions, as the considerable challenges to engineering with diode lasers have been met. At the same time the compactness and the low power consumption of diode lasers have enabled important new applications such as storing information in compact discs and DVDs, and the practical high-speed, broadband transmission of information over optical fibers, a central component of the Internet.

Construction of a Double-heterostructure Diode Laser

In addition to a means to create optical gain, a laser requires a feedback mechanism, a pair of mirrors to repeatedly circulate the light through the gain medium to build up the resulting beam by stimulated emission. The stripe structures needed to make a laser diode chip are formed on a single crystal wafer using the standard photolithographic patterning techniques of the semiconductor industry. The substrate crystal axes are first oriented relative to the patterning such that, after fabrication, a natural cleavage plane is normal to the stripe direction, and cleaving both ends of the chip provides a pair of plane, aligned crystal surfaces that act as a Fabry-Perot resonator for optical feedback. These mirrors use either the Fresnel reflectivity of the facet (often sufficient because of the high gain of diode lasers), or they can be dielectric coated to other reflectivities. This might be desired, for instance, to protect against damage from the high irradiance at the facets. This geometry gives the familiar edge-emitting diode laser.

The semiconductor crystal must be defect free to avoid scattering of carriers and of light. To grow crystal layers without defects, commercial semiconductor lasers use III-V compounds, elements taken from those columns of the periodic table. These form varying alloys with the addition of dopants that can be lattice-matched to each other and to the initial crystal substrate. The band gap of the semiconductor chosen determines the lasing wavelength region. There are three main families: GaN-based lasers with UV-blue outputs, GaAs-based lasers with red-near infrared outputs, and InP-based lasers with infrared outputs. These base crystals are precisely doped with Ga, Al, In, As, and P to precisely control the band gap and index of refraction of the layers in the diode structure.

These compounds are direct band-gap semiconductors with efficient recombination of injected holes and electrons because no phonons (lattice vibrations) are required to conserve momentum in the recombination interaction. The injection layers surrounding the junction, the cladding layers, can be indirect band-gap semiconductors (where phonons are involved).

To make a planar waveguide that concentrates the light in the junction region (confinement between the top and bottom horizontal planes of the active region in figure), the cladding layers are made of an alloy of lower refractive index (larger band gap) than the active junction region. This is then termed a double-heterostructure (DH) laser. The output power of the laser is horizontally polarized because the reflectivity of the planar waveguide is higher for the polarization direction parallel to the junction plane. Because the junction region is thin for efficient recombination (typically 0.1 μm), some light spreads into the cladding layers which are therefore made relatively thick (typically 1 μm) for adequate light confinement.

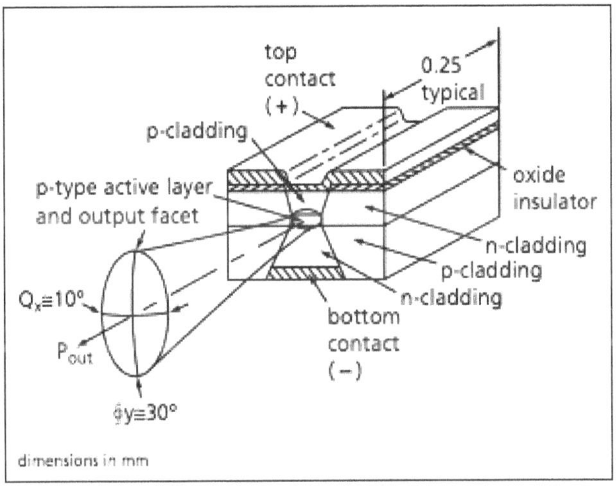

Schematic of a double heterostructure index-guided diode laser.

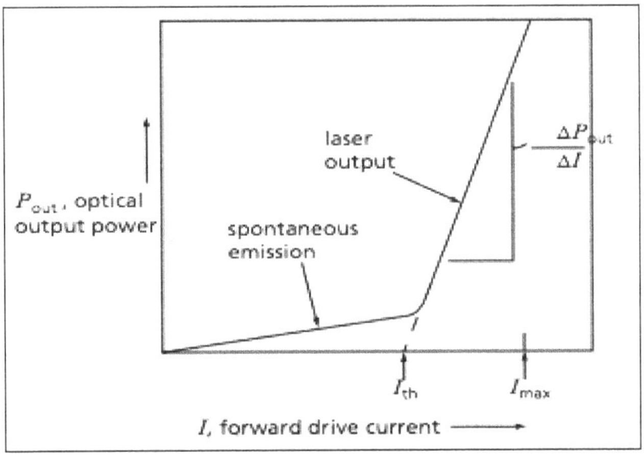

Definition of threshold current, I_{th}, and slope efficiency from the curve of light output, P_{out} vs drive current I.

Gain Guiding and Index Guiding in Diode Lasers

To confine the light laterally (between planes perpendicular to the junction plane), two main methods (with many variants) are used. The first and simplest puts a narrow conductive stripe on the p-side of the device to limit the injected current to a line, giving a gain-guided laser. There is some spreading of current under the stripe, and the light is restricted only by absorption in the unpumped regions of the junction. The transverse mode of the laser light is therefore not tightly controlled. Many high-power diode lasers, used for instance in side-pumping another solid-state laser (where mode control is less critical), are gain guided.

More efficient lateral laser mode control is achieved by fabricating, with multiple photolithographic, epitaxial, and etching steps, regions of low index of refraction on either side of the lasing stripe. This confines the light by waveguiding between planes perpendicular to the junction plane as well giving an index-guided laser. These lasers produce a stable single transverse mode of lowest order, as required in data storage applications to read compact discs, and telecommunications applications where coupling into a fiber optic is important.

Threshold Current and Slope Efficiency

Output power from a diode laser increases linearly with the drive current excess above the threshold current. This steeply rising light output curve is extrapolated backward to the zero light output intercept to define the threshold current; the weak incoherent light emission for currents below threshold is due to the spontaneous recombination of carriers such as occurs in LEDs.

When divided by the drive voltage V, the slope of the output vs current curve yields the differential (above threshold) electrical-to-optical power conversion efficiency (also termed the slope or quantum efficiency) which ranges from 50 to 80 percent for various devices.

$$slope\ efficiencency \equiv \frac{\Delta P_{out}}{V \Delta I}$$

Fabrication Methods and Quantum Wells

Three types of epitaxial crystal growth are employed in fabricating the layers of semiconductor alloys for diode laser chips: liquid phase epitaxy (LPE), metal-organic chemical vapor deposition (MOCVD), and molecular beam epitaxy (MBE).

Most early diode lasers were made by the LPE process, and it is still in use for many commercial diode lasers and LEDs. In this process, a heated, saturated solution is placed in contact with the substrate, and it is cooled, leaving an epitaxial film grown on the substrate. High-quality crystal layers are readily produced by this technique, but it is hard to control alloy composition. Furthermore, making thin layers is difficult. Because Quantum well (QW) structures, discussed below, require very thin layers. The LPE process is not appropriate for these devices; they are fabricated using the MOCVD or MBE process. In the MOCVD process, gases transport the reactants to the heated substrate, where they decompose and the epitaxial layer slowly grows. In the high-vacuum MBE process, the reactants are evaporated onto the substrate, giving a very slow, controlled epitaxial growth. The equipment for MBE is more expensive, and the process is slower making this process most suitable for critical and complex devices of low production volume. The emergence of the MOCVD and MBE processes made possible improved diode lasers employing quantum well structures as their active regions. A quantum well is a layer of semiconductor of low electron (or hole) potential energy between two other layers of higher potential energy. The well layer is made thin enough, typically less than 0.01 μm, to be comparable in size to the Bohr radius of the electron (or hole) in the material. This brings in quantum effects—the confined carrier acts, in the direction perpendicular to the layer plane, as a one-dimensional particle in a potential well. In practical terms, the density of carriers is greatly increased in this QW structure, and the laser threshold current decreases by an order of magnitude. The laser's active region is effectively an engineered, man-made material whose properties can be designed.

Disadvantage of QW Lasers

The active region is too thin to make a reasonable waveguide. This problem is solved by inserting intermediate layers of graded index between the QW and both cladding layers. This is termed the graded-index separate confinement heterostructure (GRINSCH) since the carriers are confined to the QW while the laser mode is confined by the surrounding layers. The electrical and optical confinements are separate. For higher output power, several QWs separated by buffer layers can be stacked on top of one another,—a multiple quantum well (MQW) structure. A structure with only one quantum layer is designated a single quantum well (SQL) to distinguish it from a MQW.

The lasing wavelength in QW lasers is determined by both the bulk band gap and the first quantized energy levels; it can be tuned by varying the QW thickness. Further adjustment of the wavelength is possible with strained quantum QWs. If an epitaxial layer is kept below a critical thickness, an alloy with a lattice mismatch to the substrate will distort its lattice (in the direction normal to the substrate) to match the substrate lattice instead of causing misfit dislocations. The strain in the lattice of the resulting QW changes its band gap, an effect taken advantage of to put the lasing wavelength into a desired region.

Wavelength stabilization with distributed and surface-emitting output geometries. The wavelength of a AlGaAs diode laser tunes with substrate temperature at a rate of about 0.07 nm/°C, a rapid enough rate that many applications require the baseplate of the device to be mounted on a temperature controlled thermoelectric cooler to maintain wavelength stability. Wavelength, threshold current, and efficiency are all sensitive to changes in temperature. If the laser baseplate temperature is allowed to drift, in addition to this long-term shift in wavelength, the output oscillation will jump between drifting longitudinal cavity modes and thus exhibit small, rapid, discontinuous changes in wavelength and output power which often are undesirable. To address this issue, gratings are fabricated into the laser, either at the ends of the gain stripe to create a distributed Bragg reflector (DBR) structure, or along the whole length of the gain region to create a distributed feedback (DFB) structure. The grating has a period on the order of 200 nm and is fabricated using interferometric techniques. (The beam from an argon or HeCd laser is split into two; the beams are are then overlapped to create fringes, which in turn are used to expose the photoresist in the photolythography process.) The gratings work by providing a small reflected feedback at each index step. The single frequency whose multiple fed-back reflections add up in phase determines the lasing wavelength and stabilizes it against changes in drive current and baseplate temperature. Because the laser operates in a single frequency, noise is also reduced. DBR and DFB lasers are used extensively as telecommunication light sources.

The DFB laser is an edge emitter. In the second-order gratings fabricated in both DFB and DBR lasers, the first-order diffraction is perpendicular to the surface of the grating. By providing an output window on one of the gratings in a DBR laser, the output can be brought out through the surface of the chip, i.e., a surface-emitting laser. Recently, another surface-emitting structure, the verticalcavity surface-emitting laser (VCSEL), has come into use in telecommunication links. In this structure multilayer mirrors are fabricated on the top and bottom of the QW gain region to give feedback. Consequently, the laser output is perpendicular to the active QW plane.

The epitaxial growth process of this structure is more difficult than that for edge emitters This is because provision must be made to channel current flow around the mirrors to reduce device resistance and because precise control of the mirror layer thicknesses is needed to locate standing wave peaks at the QW active layer(s). Countering these drawbacks, by having no facets to cleave, these lasers have a similar topology to LEDs. They can be tested at the wafer level and packaged using similar low-cost manufacturing methods. In addition, VCSELs have large-area circular beams (defined by the circular limiting aperture of the mirrors) and low threshold currents so they couple well into optical fibers and fit well in low-power (~1 mW) communication system applications.

Diode Laser Beam Conditioning

Because the emitting aperture is small on a typical diode laser, beam divergences are large. For example, the emitting area for the index-guided laser shown in figure might be as small as 3×1 μm, resulting in diverges of 10×30 degrees. The optics needed to collimate this beam or to focus it into a fiber must work at a high numerical aperture, resulting in potential lens aberrations, and requiring critical focusing because of short depth of field. Focal lengths must be kept short as well or the optics rapidly become large. The beam inself is elliptical and may be astigmatic. It is often desireable to first circularize the beam spot with an anamorphic prism pair or cylinder lens before coupling the laser output into an optical train. Higher-power lasers with high-order modes cause additional problems when coupling their beams into a fiber or optical system.

A wide variety of specialized components are available to address these issues, from molded miniature aspherical lenses to hyperbolic profile fiber cylinder lenses, but all require critical focusing adjustment in their mounting into the laser diode housing. For these reasons many diode lasers are offered with beamcorrecting optics built in by the manufacturer who has the appropriate tooling for the task. Typically these lasers are available as collimated units, or as fiber-coupled ("pigtailed") devices.

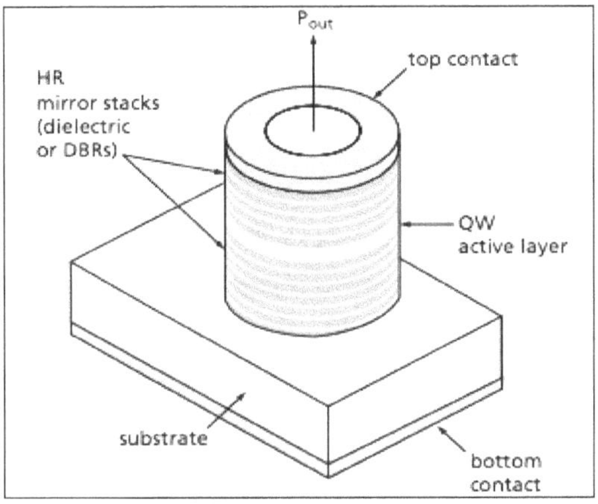

Schematic of the VCSEL structure, with light emitted perpendicular to the active layer.

High-power Diode Lasers

Single transverse mode diode lasers are limited to 200 mW or less of output power by their small emitting aperture. The facet area is so small (about 3×1 μm) that this power still represents a high irradiance (~7 MW/cm^2). The output is limited to this level to stay safely under the irradiance that would cause damage to the facet.

Enlarging the emitting area with an increase of the lateral width of the active stripe is the most common method of increasing the laser output power, but this also relaxes the single transverse mode constraint. Multiple transverse lateral modes, filaments, and lateral mode instabilities arise as the stripe width increases. For example, in a GaAs laser running at 808 nm, the output power rises linearly from 500 mW to 4 W as the lateral width of the emitting aperture increases from 50 to 500 μm. Howerer, the M2 value of the beam in this plane increases from 22 to 210. The M^2 increase makes it difficult to couple these devices to fibers, but they find considerable application in pumping solid-state laser chips designed to accept a high-numerical-aperture focus.

The pump diode lasers for even higher-power DPSS lasers are made as linear arrays of 20 or more stripe emitters integrated side by side on a 1-cm-long semiconductor bar. The bar is mounted in a water-cooled housing to handle the heat load from the high drive current. These diode laser arrays provide from 20 to 40 W of continuous output power at wavelengths matching the absorption bands of different laser crystals (e.g., 808 nm for pumping Nd:YAG lasers). The individual stripe emissions are not coherently related, but bars can be used to side pump a laser rod, just as the arc lamps they replace formerly did. Another common delivery geometry is a bundle of multimode fibers, fanned into a line of fibers on one end with each fiber butted against an individual stripe on the bar, with the other end of the bundle gathered into a circular grouping. This converts the bar output into a round spot focusable onto the end of the crystal to be pumped. Finally, for even more output, a few to a dozen bars are mounted like a deck of cards one on top of another in a water-cooled package, connected in series electrically, and sold as a stacked array. These can deliver in excess of 500W output power from one device.

Packaging, Power Supplies and Reliability

For low-power lasers, the industry uses standard semiconductor device package designs, hermetically sealed with an output window. Lasers with higher power dissipation come with a copper baseplate for attachment to a finned heat sink or thermoelectric cooler (TEC). Many are offered coupled into a fiber at the manufacturing plant in a pigtailed package (with an output fiber attached) because of the criticality in mounting the coupling optics as mentioned above.

Careful heat sinking is very important because all the major device parameters—wavelength, threshold current, slope efficiency, and lifetime—depend on device temperature (the cooler, the better). Temperature-servoed TECs are preferred for stable operation with the temperature sensor for feedback mounted close to the diode laser.

Diode lasers are susceptible to permanent damage from static electricity discharges or indeed any voltage transient. Their low operating voltage (~2 V) and ability to respond at high speed means that a static discharge transient can be a drive current spike above the maximum safe level and result in catastrophic facet damage. All the usual antistatic electricity precautions should be taken in working with diode lasers: cotton gloves, conductive gowns, grounded wrist straps, work tables, soldering irons, and so on. Correspondingly, the drive current power supply should be filtered against surges and include "slow starting" circuitry to avoid transients. Diode lasers degrade with high power and long operating hours as crystal defects migrate and grow, causing dark lines or spots in the output mode pattern and increases in threshold current or decreases in slope efficiency. The best way to prolong life is to keep the laser baseplate running cool. Remember that accelerated life tests are run by operating at high baseplate temperature. Expectations for the median life of a device are set from measurements of large populations—individual devices can still suddenly fail. Nevertheless, the industry expectations today for standard diode lasers run within their ratings is ~105 hours of operation for low-power diode lasers and perhaps an order of magnitude less for the high-power versions.

The applications mentioned in the discussion above, and a few others, are summarized in the following table and ordered by wavelength. The newer GaN lasers provide low power (10–100 mW) blue and UV wavelengths finding applications as excitation sources for biomedical fluorescence studies (DNA sequencing, confocal microscopy). The dominant application for diode lasers is as readouts for optical data storage, followed by growing numbers in use in telecommunications. For high-power (>1 W) diode lasers, the main application is as optical pumps for other solid state lasers.

Optically Pumped Lasers

Optically pumped lasers use photons of light to directly pump the lasing medium to the upper energy levels. The very first laser, based on a synthetic ruby crystal, was optically pumped. Optically pumped lasers can be separated into two broad categories: lamp-pumped and laser pumped. In a lamp-pumped laser, the lasing medium, usually a solid-state crystal, is placed near a high-intensity lamp and the two are surrounded by an elliptical reflector that focuses the light from the lamp into the crystal.

In general, ignoring the efficiency of the pump laser itself, laser pumping is a much more efficient mechanism than lamp pumping because the wavelength of the pump laser can be closely matched to specific absorption bands of the lasing medium, whereas most of the light from a broad-spectrum lamp is not usefully absorbed in the gain medium and merely results in heat that must be removed from the system. Furthermore, the size of the laser pump beam can be tightly controlled, serving as a gain aperture for improving the output mode characteristics of the pumped laser medium. On the other hand, laser pumping is often not suitable for high-energy applications where large laser crystals are required.

Diode Laser Applications

Wavelength λ (nm)	Lattice Material	Application
375	GaN	Biomedical fluorescence
405	GaN	Biomedical fluorescence, DVD mastering
440	GaN	Biomedical fluorescence, HeCd laser replacement
473	GaN	Biomedical fluorescence
635	GaAs	Pointers, alignment, HeNe laser replacement
650	GaAs	DVD readouts
670	GaAs	Barcode scanners, pointers, alignment
780	GaAs	Audio CD readouts
785	GaAs	Raman spectroscopy
808	GaAs	Optical pumps for Nd:YAG lasers, thermal printing
940	InP	Optical pumps for Yb:YAG lasers
980	InP	Optical pumps for Er fiber telecom amplifiers
1310	InP	Input source for telecom short-wavelength channels

1455	InP	Optical pump for Raman gain in standard telecom fiber
1550	InP	Input source for telecom long-wavelength channels

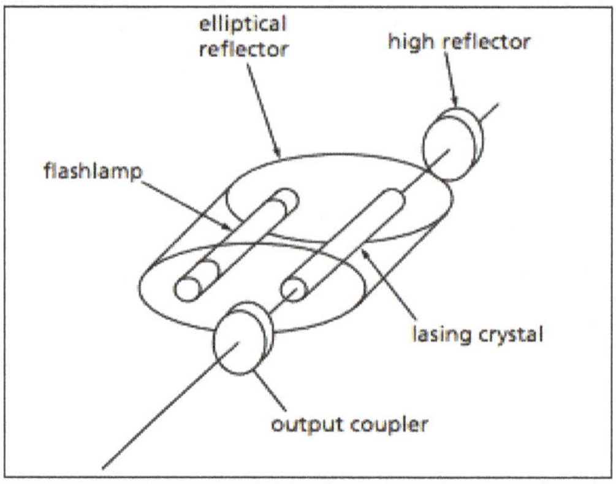

Schematic of a lamp-pumped laser.

Diode-pumped Solid State Lasers

DPSS Laser Revolution

The optical difficulties encountered with diode lasers—difficulty in coupling to the high divergence light, poor mode quality in the slow axis of wide-stripe lasers, low output power from single-transverse-mode lasers— led to a new philosophy about how best to use these efficient, long-lived, compact light sources.

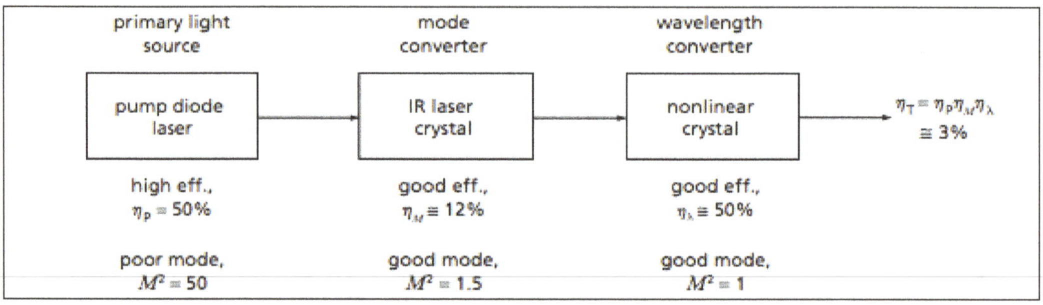

The logic for DPSS lasers.

The logic is simple. The primary light source (the diode laser) pumps another laser (an infrared crystal laser) to convert to a good mode, the beam of which is wavelength converted (by nonlinear optics techniques) to a visible output. The diode laser source replaces the discharge lamp for optically pumping the gain crystal in a traditional, high-efficiency, infrared laser. The infrared beam is generated in that independent resonator with a good mode, and consequently it can be efficiently converted with an intracavity nonlinear crystal to a visible beam with a good mode. Though power is lost at each step, the result is still a single-mode visible beam generated with a total electrical-to-optical conversion efficiency of several percent. These DPSS lasers are replacing the older visible gas lasers whose conversion efficiencies rarely reach 0.1 percent.

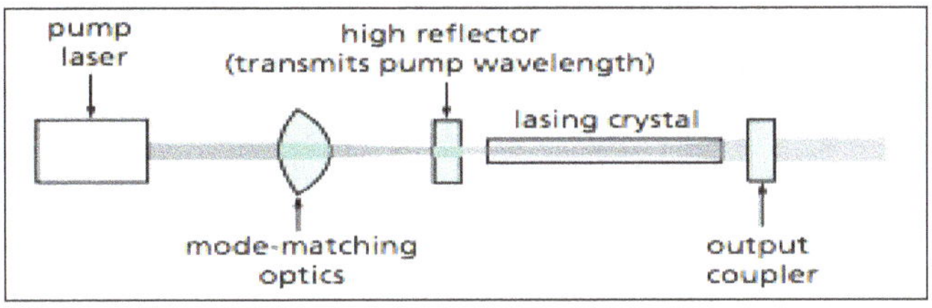

Schematic of a laser-pumped laser.

End- and Side-pumping Geometries

The first DPSS lasers were made by focusing the diode light from a single laser diode emitter through the high-reflector coating (at 1.06 μm) on the end of the Nd:YAG rod. This "end-pumping" geometry provide good overlap between the pumped volume and the lasing volume, but it limited the pump power to that available from single-mode diode emitters. In order to increase laser output and reduce cost (diode lasers suitable for end pumping are twice as expensive as diode laser arrays), diode arrays were mounted along the length of the laser rod. However, because of poor overlap of the pump beam with the 1.06 μm beam, the efficiency of this "side-pumping" technique was only half that of end-pumping geometries. No pump diode cost savings resulted. Then in the late 1980s two advances were made. First, a variety of new laser rod materials, better tailored to take advantage of diode laser pumps, were introduced. Nd:YVO4 crystals have five times the gain cross section of Nd:YAG, and the Nd can be doped into this crystal at much higher concentrations. This decreases the absorption depths in the crystal from cm to mm, easing the collimation or focusing quality required of the pump beam. This crystal had been known, but could be grown only to small dimensions, which is acceptable for diode-pumped crystals. Another crystal introduced was Yb:YAG, pumped at 980 nm and lasing at 1.03 μm— leaving very little residual heat in the crystal per optical pumping cycle and allowing small chips of this material to be pumped at high levels. Second, means were devised to make micro-cylindrical lenses (focal lengths less than a mm) with the correct surfaces (one type is a hyperbolic profile) for collimating or reducing the fast-axis divergence of the diode laser output. With good tooling and beam characterization these are correctly positioned in the diode beam and bonded in place to the diode housing. This allows more conventional lenses, of smaller numerical aperture, to be used in subsequent pump light manipulations.

End-pumping with Bars

With these two new degrees of freedom, laser designers realized they could create optical trains that would give them end-pumping system efficiencies (achieve good overlap between pump and lasing modes) with diode arrays as pump sources to obtain a lower diode cost per watt in their systems. This produced an explosion of unique DPSS laser designs generically described as "end-pumping with bars." Figure shows the example previously mentioned, delivering the array light through a fiber bundle, with the fibers at one end spread out to butt align with the linear stripes of an array, and the other end of the bundle gathered to an approximately circular spot. Although the circular spot is large, its focal image, formed with high numerical aperture (NA) optics, is small enough to satisfactorily overlap the IR cavity laser mode. The small depth of focus, from the high NA optics, is inconsequential here because of the short absorption depth in the Nd:YVO4 laser crystal. The laser head can be disconnected from the diode modules at the fiber coupler without loss of alignment. In another example, an even higher-NA optic (comprising a cylinder lens and a molded aspheric lens) was used to directly focus the 1-cm width of a micro-lensed array bar onto the end of a Nd:YVO4 gain crystal. This produced an oblong pump spot, but good overlap with the IR cavity mode was achieved by altering the infrared cavity (inserting two intracavity beam expansion prisms in that arm) to produce a 5:1 elliptical cavity mode in the gain crystal. Another design used a nonimaging pyramidal "lens duct" to bring in the pump light from a diode laser stack to the end of a gain crystal. Yet another brought light from several arrays into a lasing rod centered in a diffuse-reflecting cavity by means of several planar (glass-slide) waveguides, each piercing a different sector of the reflector sidewall. These are but a few of the design approaches that have been successfully taken.

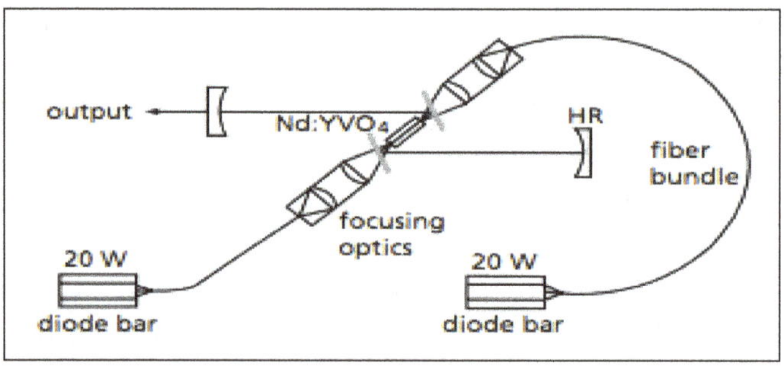

Schematic of an "end-pumping with bars" geometry using fiber bundle delivery, one of many variants on the DPSS laser theme.

Microchip Lasers

Another procedure that can be used to make available potentially inexpensive, mass produced, low power, visible output DPSS lasers is mimicking semiconductor chip processing methods. In the late 1980s, MIT Lincoln Labs took this approach and created the "microchip" laser. A thin Nd:YVO4 plate is polished flat and then diced into ~2-mm-square chips. Each of these chips is then optically contacted to similar, flat, diced KTP doubling crystal plates to make a cube. Prior to dicing, the surfaces that will become the outer cube surfaces are coated for high reflectivity at 1.06 mm. When single-mode diode laser pump light is focused through one mirrored end of the cube, the heat produced makes a thermally-induced waveguide that creates a stable cavity for IR lasing. Since the KTP crystal is within this cavity, the IR lasing is converted to a 532-nm (green) output beam with 10's of milliwats of output. The diode temperature must be controlled to maintain a stable pump wavelength and thermal waveguide. In addition, the cube temperature should be stabilized. Because of the short cavity, the IR laser operates at a single longitudinal mode, and the cavity length must be thermally tuned to keep the mode at the peak of the gain curve.

Green Noise Problem

As the early DPSS laser designs giving visible-output beams were being introduced, it became apparent that there was a problem unique to this architecture. The visible output power, 532 nm in the green spectrum, could break into high-frequency chaotic oscillations of nearly 100-percent peak-to-peak amplitude. This was named the "green problem" by Tom Baer (then at Spectra-Physics), who in 1986 showed the effect to be due to the dominance of sum-frequency-mixing (SFM) terms coupling different longitudinal modes over second-harmonic-generation (SHG) terms, in the nonlinear conversion step from IR to visible output. Several conditions (all met by the new laser designs) lead to this effect: (1) the IR laser cavity is short (~10 cm or less) with only a few longitudinal modes oscillating, (2) nonlinear conversion efficiencies are high (20% or more), and (3) nonlinear phase-matching bandwidths span several longitudinal mode spacings (true of the commonly used KTP doubling crystal). Then the sum frequency mixing output losses couple the longitudinal modes in relaxation oscillations where the turn on of one mode turns off another.

Two early solutions to this problem emerged. The first is to make the IR cavity long enough to give hundreds of oscillating modes, so that the noise terms average to insignificance as in a long gas laser. The second is to make the IR oscillation run on a single longitudinal mode so that there are no SFM terms. This can be done by using intracavity frequency control elements such as an etalon, or by using a ring cavity (with a Faraday-effect biasing element to maintain the direction of light travel around the ring). Ring cavities eliminate the standing-wave interference effect of linear cavities, termed "spatial hole burning," and the laser runs single frequency when this is done. As more experience was gained with DPSS laser design, other clever solutions to the "green problem" were found, tailored to each particular device and often held as trade secrets. It can be surmised that these involve precise control of wavelength, spatial hole burning, beam polarization, and cavity-element optical path differences to reduce the strength of longitudinal mode SFM terms.

Uniqueness of DPSS Laser Designs and Laser Reliability

Unlike the gas lasers they replace, no universal approach is applied in the details of different DPSS laser designs and laser models. There is a large variety of solutions to the major problems, many solutions are unique, and many are held as proprietary. Major design differences are found in:

- The means for optically coupling the pumping light into the gain medium.
- The management of the thermal lens produced by absorption of the pump light in the cavity.
- The control of green noise.
- The strain-free mounting, heat sinking, and placement of the small lasing and nonlinear crystals in the laser cavity.

- The hermetic sealing of the laser cavity to protect the often delicate crystals and critical alignments of components.

The intracavity space must be hermetically sealed there usually is no field repair, maintenance, or adjustment of a DPSS laser head. If it fails, it is returned to the manufacturer.

It is evident that DPSS lasers are a lot less generic than the gas lasers they replace. For a problem with a particular laser model, there may be no standard solution available in the technical literature. With so many variables, there often are surprises when new designs are first manufactured and introduced. Under these circumstances, the user is advised to pick a supplier with a record of years of consistent manufacture, who has over time dealt with his own unique set of component and assembly problems. If this advice is followed, then the expectation with current products is that a new DPSS laser will operate reliably for 10,000 hours or more.

An example of a DPS laser product line— The melles griot visible output lasers.

Figure depicts the mix of laser crystals, laser operating wavelengths, and doubling crystals generating the four visible output wavelengths of the present Melles Griot product line of continuous wave DPSS lasers.

Other Notable DPSS Lasers

The Er-doped fiber amplifier (EDFA) is not a laser, but it is an optically pumped amplifier for the 1550-nm long-wavelength long-haul fiberoptic channels that make the worldwide web possible. Pumping an Er-doped silica fiber with 980-nm diode laser light inverts the populations of energy levels in the Er ions to provide gain for optical telecommunication signals run through the same fiber. This optical amplifier is much simpler than the discrete electronic repeaters it replaced. A Lucent Technologies executive expressed the importance of this when he said: "What broke [wavelength division multiplexing telecommunications] free was the invention of the [EDFA] optical amplifier."

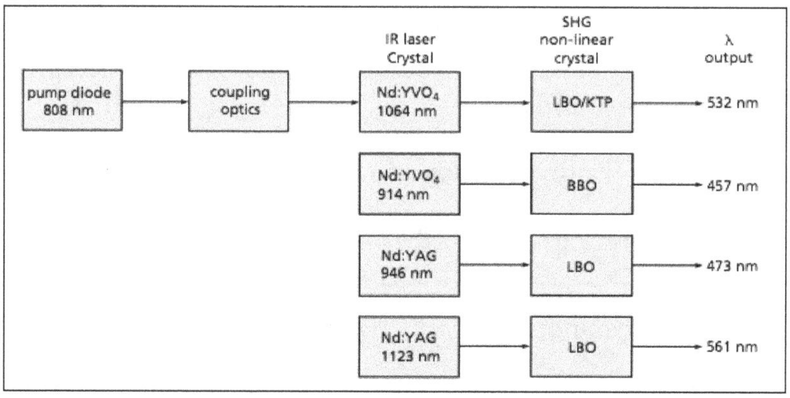

Melles Griot DPSS laser optical trains for producing four different visible output wavelengths.

The Q-switched industrial DPSS laseris a 1-W-average-power, ultraviolet (355 nm), high-repetition-rate (30 kHz) system. Output is obtained by doubling the 1.064-μm Q-switched fundamental to green at 532 nm, and then mixing the green beam with the residual transmitted infrared to 355 nm. This process is straightforward in a high-peak-power pulsed beam—just a matter of inserting the appropriate doubling and tripling crystals. What is remarkable is that DPSS laser designs have matured sufficiently to make this possible in a hands-off, long lived, system rugged enough to survive and be useful in an industrial environment.

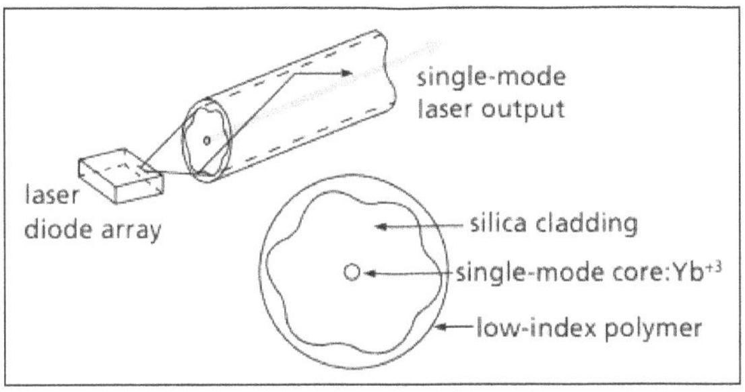

Schematic diagram of the structure of a double-clad fiber, and the method of pumping the inner core by direct illumination into the large diameter of the outer cladding.

The double-clad fiber laseris shown in figure. Fiber lasers work by optically pumping (with a diode laser) a doped fiber and adding mirrors for feedback at either end of the fiber. In the dual-clad fiber, the Yb-doped single-mode fiber core is surrounded by a large diameter cladding (with a corrugated star-shaped cross section in the figure) that is itself clad by a low-index polymer coating. Diode laser light at 940 nm is readily launched into and guided in the large diameter outer cladding, and the corrugated cross-section of this fiber suppresses the helical ray modes of propagation that would have poor overlap with the inner core. Over the length of the fiber, the pump light is absorbed by the single-mode core, and high- power lasing near 1.03 μm in a low- order mode is produced. The quantum efficiency of the Yb lasing cycle (ratio of pump wavelength to lasing wavelength) is 91 percent, which leaves little heat deposited in the fiber. Over 1 kW of output at 80-percent slope efficiency has been produced in such a fiber laser.

Laser Physics 2

- **Gain Switching**
- **Mode-locking**
- **Q-switching**
- **Laser Pumping**
- **Active Laser Medium**

The branch of optics that describes the theories and practices of lasers is known as laser physics. Gain switching, mode-locking, Q-switching and laser pumping are some of the techniques used in laser physics. This chapter discusses in detail these techniques and methods related to laser physics.

Laser physics is a branch of physics where the fundamental and the applied aspects of laser science are studied. Basically laser physics is a part of optics which deals with the laser theory.

The spatial coherence helps the laser beam to stay narrow over a vast region. This narrowness is helpful for the laser pointer. In this case, the laser beam achieves a very high irradiance and the divergence of light is very low.

Again, laser is also having a high temporal coherence which helps laser beam to give a very narrow spectrum. This temporal coherence laser beam is polarized one.

Laser physics is used to manipulate the single quantum particles for implementation quantum information processing.

To analysis the interaction between light and the matters, laser physics is useful. The high intensity light matter interaction can be measured by the ultrafast optical laser.

Again, to create an exotic quantum state like Bose Einstein condensation state, laser is very useful. Laser physics is also using to determine the quantum mechanical superposition and quantum entanglement information.

In atomics physics, there is also much application of laser physics, like the interaction between the atoms and the photons at the single particle level can be determined.

There are many applications of laser physics in various technologies, like fiber optics technology, optical disk drives, laser printers, barcode scanner, DNA sequencing instrument, laser surgery, skin treatment, non linear optics, ultra short pulses, optical communication, general optics, optoelectronics, and quantum optics.

Gain Switching

Gain switching is a method for pulse generation by quickly modulating the laser gain via the pump power. If a high pump power is suddenly applied to a laser, laser emission sets in only with a certain delay, as it starts with weak fluorescence light, which first needs to be amplified in a number of resonator round trips. Therefore, some amount of energy can be stored in the gain medium, which is subsequently extracted in the form of a short pulse. The pulse obtained can be shorter than the pump pulse and also shorter than the upper-state lifetime; the dynamics are essentially as in the phenomenon of spiking, where the pump power is applied for a short enough time to generate only a first spike.

The higher the pump pulse energy, the shorter is the pulse build-up time, and consequently the required pump pulse duration. The pulse build-up time of a gain-switched laser can be increased by using a longer laser resonator, but this also increases the output pulse duration.

The pump power may be completely switched off between the pulses, or may be kept at a level just below the laser threshold.

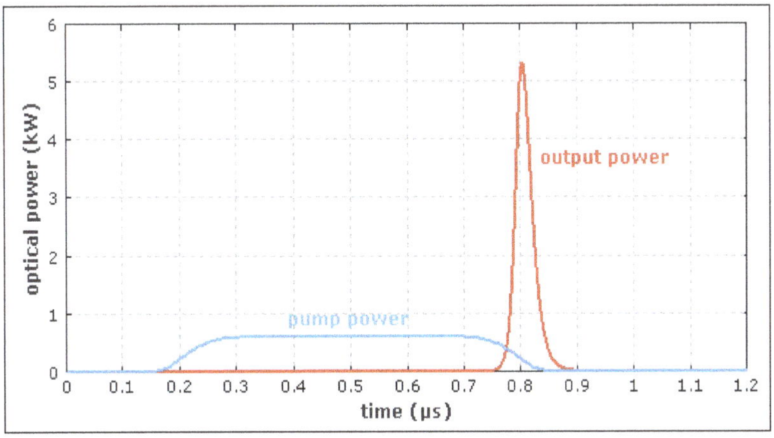

Simulated power evolution for gain switching of a solid-state laser. The pump energy is so high that the laser pulse is emitted at the end of the pump pulse. For a higher pump energy, the laser pulse would start too early, and a second pulse may be emitted.

Application of Gain Switching

Gain switching can be applied to various types of lasers:

Gas Lasers

Some gas lasers can be operated with a pulsed excitation current. This is often done with TEA carbon dioxide lasers (CO_2 lasers), which are suitable only for pulsed operation. However, the result is not necessarily gain switching in the sense of using the first emitted spike; it may also be a kind of quasi-continuous-wave operation over a limited duration.

Dye Lasers

In other cases, a pulsed laser is used for optically pumping a gain-switched laser. For example, nitrogen ultraviolet lasers or excimer lasers can be used for pulsed pumping of dye lasers.

Solid-state Lasers

A solid-state bulk laser or fiber laser can emit a single spike if it is pumped only for up to a few microseconds. The pulse duration may be of the order of tens to hundreds of nanoseconds. That operation mode is rarely used, however, due to the limited pulse energy.

Laser Diodes

It is also possible to operate a laser diode with short current pulses, or with a continuously modulated signal. This can lead to pulses with durations of a few nanosecond or even down to a few tens of picoseconds, and with pulse repetition rates up to several gigahertz, as used in telecom applications. With additional pulse compression, one can reduce the pulse durations further to below 4 ps.

In contrast to mode-locked lasers, gain-switched laser diodes easily allow one to adjust the pulse repetition rate in a wide range, since it can be controlled with an electronic driver without changing the laser resonator setup. Also, they are simpler and more compact. However, the timing jitter and also fluctuations of other pulse parameters are larger than for a mode-locked laser. Subsequent pulses are not mutually coherent.

The pulse energy of a gain-switched laser diode is fairly limited, since the peak power can usually not be much higher than for longer pulses. When trying to produce longer pulses with higher energy, the gain-switched peak can be disturbing, if one needs a smooth temporal pulse profile.

An interesting approach to the generation of energetic nanosecond pulses is to combine a gain-switched laser diode with a fiber amplifier. In this way, pulses with microjoule or millijoule energies can be generated. This method provides a high flexibility in terms of pulse duration, shape and repetition rate. However, a rather high amplifier gain is needed due to the low seed pulse energy, and therefore one often requires more than one amplifier stage and additional optical filters for achieving sufficient suppression of amplified spontaneous emission.

Gain-switched laser diodes may also be used as seed lasers for optical parametric amplifiers.

Mode-locking

Mode-locking is a technique in optics by which a laser can be made to produce pulses of light of extremely short duration, on the order of picoseconds (10^{-12} s) or femtoseconds (10^{-15} s). A laser operated in this way is sometimes referred to as a femtosecond laser, for example in modern refractive surgery. The basis of the technique is to induce a fixed-phase relationship between the longitudinal modes of the laser's resonant cavity. Constructive interference between these modes can cause the laser light to be produced as a train of pulses. The laser is then said to be 'phase-locked' or 'mode-locked'.

Laser Cavity Modes

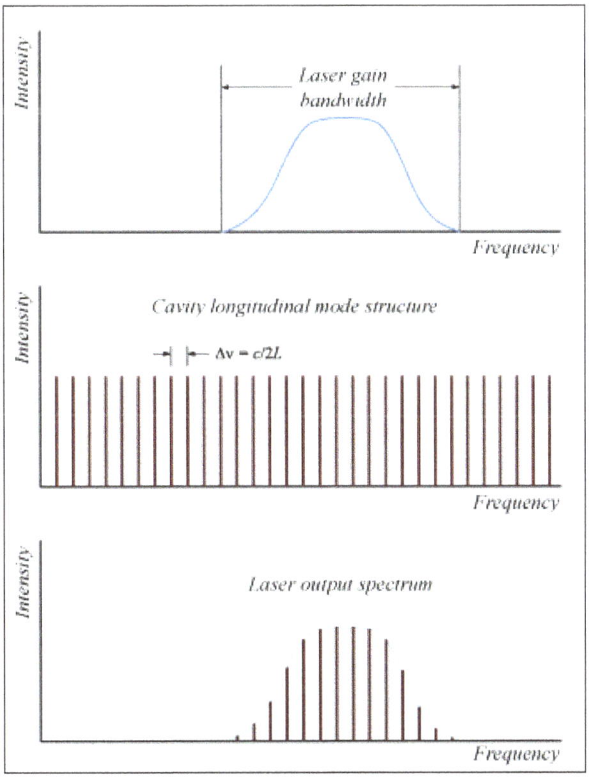

Laser Mode Structure.

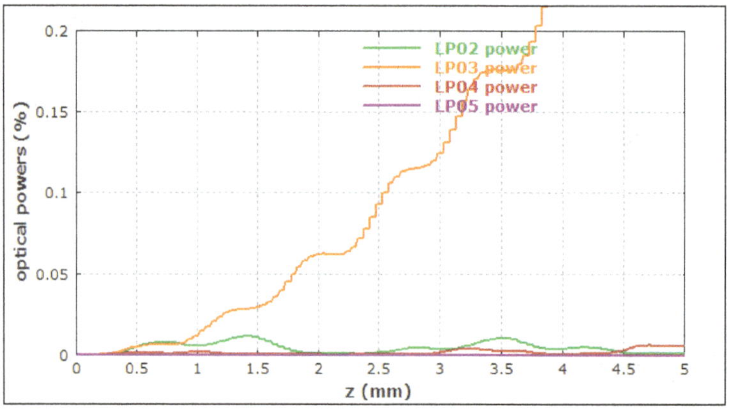

Although laser light is perhaps the purest form of light, it is not of a single, pure frequency or wavelength. All lasers produce light over some natural bandwidth or range of frequencies. A laser's bandwidth of operation is determined primarily by the gain medium from which the laser is constructed, and the range of frequencies over which a laser may operate is known as the gain bandwidth. For example, a typical helium–neon laser has a gain bandwidth of about 1.5 GHz (a wavelength range of about 0.002 nm at a central wavelength of 633 nm), whereas a titanium-doped sapphire (Ti:sapphire) solid-state laser has a bandwidth of about 128 THz (a 300-nm wavelength range centered at 800 nm).

The second factor to determine a laser's emission frequencies is the optical cavity (or resonant cavity) of the laser. In the simplest case, this consists of two plane (flat) mirrors facing each other, surrounding the gain medium of the laser (this arrangement is known as a Fabry–Pérot cavity). Since light is a wave, when bouncing between the mirrors of the cavity, the light will constructively and destructively interfere with itself, leading to the formation of standing waves or modes between the mirrors. These standing waves form a discrete set of frequencies, known as the *longitudinal modes* of the cavity. These modes are the only frequencies of light which are self-regenerating and allowed to oscillate by the resonant cavity; all other frequencies of light are suppressed by destructive interference. For a simple plane-mirror cavity, the allowed modes are those for which the separation distance of the mirrors L is an exact multiple of half the wavelength of the light λ, such that $L = q\lambda/2$, where q is an integer known as the mode order.

In practice, L is usually much greater than λ, so the relevant values of q are large (around 10^5 to 10^6). Of more interest is the frequency separation between any two adjacent modes q and q+1; this is given (for an empty linear resonator of length L) by $\Delta\nu$:

$$\Delta t = \frac{0.441}{N\Delta\nu}.$$

Where, c is the speed of light ($\approx 3\times10^8$ m·s^{-1}).

Using the above equation, a small laser with a mirror separation of 30 cm has a frequency separation between longitudinal modes of 0.5 GHz. Thus for the two lasers referenced above, with a 30-cm cavity, the 1.5 GHz bandwidth of the HeNe laser would support up to 3 longitudinal modes, whereas the 128 THz bandwidth of the Ti:sapphire laser could support approximately 250,000 modes. When more than one longitudinal mode is excited, the laser is said to be in "multi-mode" operation. When only one longitudinal mode is excited, the laser is said to be in "single-mode" operation.

Each individual longitudinal mode has some bandwidth or narrow range of frequencies over which it operates, but typically this bandwidth, determined by the Q factor of the cavity, is much smaller than the intermode frequency separation.

Mode-locking Theory

In a simple laser, each of these modes oscillates independently, with no fixed relationship between each other, in essence like a set of independent lasers all emitting light at slightly different frequencies. The individual phase of the light waves in each mode is not fixed, and may vary randomly due to such things as thermal changes in materials of the laser. In lasers with only a few oscillating modes, interference between the modes can cause beating effects in the laser output, leading to fluctuations in intensity; in lasers with many thousands of modes, these interference effects tend to average to a near-constant output intensity.

If instead of oscillating independently, each mode operates with a fixed phase between it and the other modes, the laser output behaves quite differently. Instead of a random or constant output intensity, the modes of the laser will periodically all constructively interfere with one another, producing an intense burst or pulse of light. Such a laser is said to be 'mode-locked' or 'phase-locked'. These pulses occur separated in time by $\tau = 2L/c$, where τ is the time taken for the light to make exactly one round trip of the laser cavity. This time corresponds to a frequency exactly equal to the mode spacing of the laser, $\Delta\nu = 1/\tau$.

The duration of each pulse of light is determined by the number of modes which are oscillating in phase (in a real laser, it is not necessarily true that all of the laser's modes will be phase-locked). If there are N modes locked with a frequency separation $\Delta\nu$, the overall mode-locked bandwidth is $N\Delta\nu$, and the wider this bandwidth, the shorter the pulse duration from the laser. In practice, the actual pulse duration is determined by the shape of each pulse, which is in turn determined by the exact amplitude and phase relationship of each longitudinal mode. For example, for a laser producing pulses with a Gaussian temporal shape, the minimum possible pulse duration Δt is given by,

$$\Delta t = \frac{0.441}{N\Delta\nu}.$$

The value 0.441 is known as the 'time-bandwidth product' of the pulse, and varies depending on the pulse shape. For ultrashort pulse lasers, a hyperbolic-secant-squared (sech^2) pulse shape is often assumed, giving a time-bandwidth product of 0.315.

Using this equation, the minimum pulse duration can be calculated consistent with the measured laser spectral width. For the HeNe laser with a 1.5-GHz spectral width, the shortest Gaussian pulse consistent with this spectral width would be around 300 picoseconds; for the 128-THz bandwidth Ti:sapphire laser, this spectral width would be only 3.4 femtoseconds. These values represent the shortest possible Gaussian pulses consistent with the laser's linewidth; in a real mode-locked laser, the actual pulse duration depends on many other factors, such as the actual pulse shape, and the overall dispersion of the cavity.

Subsequent modulation could in principle shorten the pulse width of such a laser further; however, the measured spectral width would then be correspondingly increased.

Mode-locking Methods

Methods for producing mode-locking in a laser may be classified as either 'active' or 'passive'. Active methods typically involve using an external signal to induce a modulation of the intracavity light. Passive methods do not use an external signal, but rely on placing some element into the laser cavity which causes self-modulation of the light.

Active Mode-locking

The most common active mode-locking technique places a standing wave electro-optic modulator into the laser cavity. When driven with an electrical signal, this produces a sinusoidal amplitude modulation of the light in the cavity. Considering this in the frequency domain, if a mode has optical frequency ν, and is amplitude-modulated at a frequency f, the resulting signal has sidebands at optical frequencies $\nu - f$ and $\nu + f$. If the modulator is driven at the same frequency as the cavity-mode spacing $\Delta\nu$, then these sidebands correspond to the two cavity modes adjacent to the original mode. Since the sidebands are driven in-phase, the central mode and the adjacent modes will be phase-locked together. Further operation of the modulator on the sidebands produces phase-locking of the $\nu - 2f$ and $\nu + 2f$ modes, and so on until all modes in the gain bandwidth are locked. As said above, typical lasers are multi-mode and not seeded by a root mode. So multiple modes need to work out which phase to use. In a passive cavity with this locking applied there is no way to dump the entropy given by the original independent phases. This locking is better described as a coupling, leading to a complicated behavior and not clean pulses. The coupling is only dissipative because of the dissipative nature of the amplitude modulation. Otherwise, the phase modulation would not work.

This process can also be considered in the time domain. The amplitude modulator acts as a weak 'shutter' to the light bouncing between the mirrors of the cavity, attenuating the light when it is "closed", and letting it through when it is "open". If the modulation rate f is synchronised to the cavity round-trip time τ, then a single pulse of light will bounce back and forth in the cavity. The actual strength of the modulation does not have to be large; a modulator that attenuates 1% of the light when "closed" will mode-lock a laser, since the same part of the light is repeatedly attenuated as it traverses the cavity.

Related to this amplitude modulation (AM), active mode-locking is frequency modulation (FM) mode-locking, which uses a modulator device based on the acousto-optic effect. This device, when placed in a laser cavity and driven with an electrical signal, induces a small, sinusoidally varying frequency shift in the light passing through it. If the frequency of modulation is matched to the round-trip time of the cavity, then some light in the cavity sees repeated upshifts in frequency, and some repeated downshifts. After many repetitions, the upshifted and downshifted light is swept out of the gain bandwidth of the laser. The only light which is unaffected is that which passes through the modulator when the induced frequency shift is zero, which forms a narrow pulse of light.

The third method of active mode-locking is synchronous mode-locking, or synchronous pumping. In this, the pump source (energy source) for the laser is itself modulated, effectively turning the laser on and off to produce pulses. Typically, the pump source is itself another mode-locked laser. This technique requires accurately matching the cavity lengths of the pump laser and the driven laser.

Passive Mode-locking

Passive mode-locking techniques are those that do not require a signal external to the laser (such as the driving signal of a modulator) to produce pulses. Rather, they use the light in the cavity to cause a change in some intracavity element, which will then itself produce a change in the intracavity light. A commonly used device to achieve this is a saturable absorber.

A saturable absorber is an optical device that exhibits an intensity-dependent transmission. What this means is that the device behaves differently depending on the intensity of the light passing through it. For passive mode-locking, ideally a saturable absorber will selectively absorb low-intensity light, and transmit light which is of sufficiently high intensity. When placed in a laser cavity, a saturable absorber will attenuate low-intensity constant wave light (pulse wings). However, because of the somewhat random intensity fluctuations experienced by an un-mode-locked laser, any random, intense spike will be transmitted preferentially by the saturable absorber. As the light in the cavity oscillates, this process repeats, leading to the selective amplification of the high-intensity spikes, and the absorption of the low-intensity light. After many round trips, this leads to a train of pulses and mode-locking of the laser.

Considering this in the frequency domain, if a mode has optical frequency v, and is amplitude-modulated at a frequency nf, the resulting signal has sidebands at optical frequencies $v - nf$ and $v + nf$ and enables much stronger mode-locking for shorter pulses and more stability than active mode-locking, but has startup problems.

Saturable absorbers are commonly liquid organic dyes, but they can also be made from doped crystals and semiconductors. Semiconductor absorbers tend to exhibit very fast response times (~100 fs), which is one of the factors that determines the final duration of the pulses in a passively mode-locked laser. In a *colliding-pulse mode-locked laser* the absorber steepens the leading edge while the lasing medium steepens the trailing edge of the pulse.

There are also passive mode-locking schemes that do not rely on materials that directly display an intensity dependent absorption. In these methods, nonlinear optical effects in intracavity components are used to provide a method of selectively amplifying high-intensity light in the cavity, and attenuation of low-intensity light. One of the most successful schemes is called Kerr-lens mode-locking (KLM), also sometimes called "self mode-locking". This uses a nonlinear optical process, the optical Kerr effect, which results in high-intensity light being focussed differently from low-intensity light. By careful arrangement of an aperture in the laser cavity, this effect can be exploited to produce the equivalent of an ultra-fast response time saturable absorber.

Hybrid Mode-locking

In some semiconductor lasers a combination of the two above techniques can be used. Using a laser with a saturable absorber, and modulating the electrical injection at the same frequency the laser is locked at, the laser can be stabilized by the electrical injection. This has the advantage of stabilizing the phase noise of the laser, and can reduce the timing jitter of the pulses from the laser.

Mode Locking by Residual Cavity Fields

Coherent phase information transfer between subsequent laser pulses has also been observed from nanowire lasers. Here, the phase information has been stored in the residual photon field of coherent Rabi oscillations in the cavity. Such findings open the way to phase locking of light sources integrated onto chip-scale photonic circuits and applications, such as on-chip Ramsey comb spectroscopy.

Fourier Domain Mode Locking

Fourier domain mode locking (FDML) is a laser modelocking technique that creates a continuous wave, wavelength-swept light output. A main application for FDML lasers is optical coherence tomography.

Practical Mode-locked Lasers

In practice, a number of design considerations affect the performance of a mode-locked laser. The most important are the overall dispersion of the laser's optical resonator, which can be controlled with a prism compressor or some dispersive mirrors placed in the cavity, and optical nonlinearities. For excessive net group delay dispersion (GDD) of the laser cavity, the phase of the cavity modes can not be locked over a large bandwidth, and it will be difficult to obtain very short pulses. For a suitable combination of negative (anomalous) net GDD with the Kerr nonlinearity, soliton-like interactions may stabilize the mode-locking and help to generate shorter pulses. The shortest possible pulse duration is usually accomplished either for zero dispersion (without nonlinearities) or for some slightly negative (anomalous) dispersion (exploiting the soliton mechanism).

The shortest directly produced optical pulses are generally produced by Kerr-lens mode-locked Ti-sapphire lasers, and are around 5 femtoseconds long. Alternatively, amplified pulses of a similar duration are created through the compression of longer (e.g. 30 fs) pulses via self-phase modulation in a hollow core fibre or during filamentation. However, the minimum pulse duration is limited by the period of the carrier frequency (which is about 2.7 fs for Ti:S systems), therefore shorter pulses require moving to shorter wavelengths. Some advanced techniques (involving high harmonic generation with amplified femtosecond laser pulses) can be used to produce optical features with durations as short as 100 attoseconds in the extreme ultraviolet spectral region (i.e. <30 nm). Other achievements, important particularly for laser applications, concern the development of mode-locked lasers which can be pumped with laser diodes, can generate very high average output powers (tens of watts) in sub-picosecond pulses, or generate pulse trains with extremely high repetition rates of many GHz.

Pulse durations less than approximately 100 fs are too short to be directly measured using optoelectronic techniques (i.e. - photodiodes), and so indirect methods such as autocorrelation, frequency-resolved optical gating, spectral phase interferometry for direct electric-field reconstruction or multiphoton intrapulse interference phase scan are used.

Applications

- Nuclear fusion. (inertial confinement fusion)
- Nonlinear optics, such as second-harmonic generation, parametric down-conversion, optical parametric oscillators, and generation of Terahertz radiation.
- Optical Data Storage uses lasers, and the emerging technology of 3D optical data storage generally relies on nonlinear photochemistry. For this reason, many examples use mode-locked lasers, since they can offer a very high repetition rate of ultrashort pulses.
- Femtosecond laser nanomachining – The short pulses can be used to nanomachine in many types of materials.
- An example of pico- and femtosecond micromachining is drilling the silicon jet surface of ink jet printers.
- Two-photon microscopy.
- Corneal Surgery: Femtosecond lasers can be used to create bubbles in the cornea. A line of bubbles can be used to create a cut in the cornea, replacing the microkeratome, e.g. for the creation of a flap in LASIK surgery (this is sometimes referred to as Intralasik or all-laser surgery). Bubbles can also

be created in multiple layers so that a piece of corneal tissue between these layers can be removed (a procedure known as Small incision lenticule extraction).

- A laser technique has been developed that renders the surface of metals deep black. A femtosecond laser pulse deforms the surface of the metal forming nanostructures. The immensely increased surface area can absorb virtually all the light that falls on it thus rendering it deep black. This is one type of black gold.
- Photonic Sampling, using the high accuracy of lasers over electronic clocks to decrease the sampling error in electronic ADCs.

Q-switching

Q-switching, sometimes known as giant pulse formation or Q-spoiling, is a technique by which a laser can be made to produce a pulsed output beam. The technique allows the production of light pulses with extremely high (gigawatt) peak power, much higher than would be produced by the same laser if it were operating in a continuous wave (constant output) mode. Compared to modelocking, another technique for pulse generation with lasers, Q-switching leads to much lower pulse repetition rates, much higher pulse energies, and much longer pulse durations. The two techniques are sometimes applied together.

Q-switching was first proposed in 1958 by Gordon Gould, and independently discovered and demonstrated in 1961 or 1962 by R.W. Hellwarth and F.J. McClung using electrically switched Kerr cell shutters in a ruby laser.

Principle of Q-switching

Q-switching is achieved by putting some type of variable attenuator inside the laser's optical resonator. When the attenuator is functioning, light which leaves the gain medium does not return, and lasing cannot begin. This attenuation inside the cavity corresponds to a decrease in the *Q factor* or *quality factor* of the optical resonator. A high Q factor corresponds to low resonator losses per roundtrip, and vice versa. The variable attenuator is commonly called a "Q-switch", when used for this purpose.

Initially the laser medium is pumped while the Q-switch is set to prevent feedback of light into the gain medium (producing an optical resonator with low Q). This produces a population inversion, but laser operation cannot yet occur since there is no feedback from the resonator. Since the rate of stimulated emission is dependent on the amount of light entering the medium, the amount of energy stored in the gain medium increases as the medium is pumped. Due to losses from spontaneous emission and other processes, after a certain time the stored energy will reach some maximum level; the medium is said to be *gain saturated.* At this point, the Q-switch device is quickly changed from low to high Q, allowing feedback and the process of optical amplification by stimulated emission to begin. Because of the large amount of energy already stored in the gain medium, the intensity of light in the laser resonator builds up very quickly; this also causes the energy stored in the medium to be depleted almost as quickly. The net result is a short pulse of light output from the laser, known as a *giant pulse*, which may have a very high peak intensity.

There are two main types of Q-switching:

Active Q-switching

Here, the Q-switch is an externally controlled variable attenuator. This may be a mechanical device such as a shutter, chopper wheel, or spinning mirror/prism placed inside the cavity, or (more commonly) it may be some form of modulator such as an acousto-optic device, a magneto-optic effect device or an electro-optic device — a Pockels cell or Kerr cell. The reduction of losses (increase of Q) is triggered by an external event, typically an electrical signal. The pulse repetition rate can therefore be externally controlled. Modulators generally allow a faster transition from low to high Q, and provide better control. An additional advantage of modulators is that the rejected light may be coupled out of the cavity and can be used for something else. Alternatively, when the modulator is in its low-Q state, an externally generated beam can be coupled *into* the cavity through the modulator. This can be used to "seed" the cavity with a beam that has desired characteristics (such as transverse mode or wavelength). When the Q is raised, lasing builds up from the initial seed, producing a Q-switched pulse that has characteristics inherited from the seed.

Passive Q-switching

In this case, the Q-switch is a saturable absorber, a material whose transmission increases when the intensity of light exceeds some threshold. The material may be an ion-doped crystal like Cr:YAG, which is used for Q-switching of Nd:YAG lasers, a bleachable dye, or a passive semiconductor device. Initially, the loss of the absorber is high, but still low enough to permit some lasing once a large amount of energy is stored in the gain medium. As the laser power increases, it saturates the absorber, i.e., rapidly reduces the resonator loss, so that the power can increase even faster. Ideally, this brings the absorber into a state with low losses to allow efficient extraction of the stored energy by the laser pulse. After the pulse, the absorber recovers to its high-loss state before the gain recovers, so that the next pulse is delayed until the energy in the gain medium is fully replenished. The pulse repetition rate can only indirectly be controlled, e.g. by varying the laser's pump power and the amount of saturable absorber in the cavity. Direct control of the repetition rate can be achieved by using a pulsed pump source as well as passive Q-switching.

Variants

Jitter can be reduced by not reducing the Q by as much, so that a small amount of light can still circulate in the cavity. This provides a "seed" of light that can aid in the buildup of the next Q-switched pulse.

With cavity dumping, the cavity end mirrors are 100% reflective, so that no output beam is produced when the Q is high. Instead, the Q-switch is used to "dump" the beam out of the cavity after a time delay. The cavity Q goes from low to high to start the laser build-up, and then goes from high to low to "dump" the beam from the cavity all at once. This produces a shorter output pulse than regular Q-switching. Electro-optic modulators are normally used for this, since they can easily be made to function as a near-perfect beam "switch" to couple the beam out of the cavity. The modulator that dumps the beam may be the same modulator that Q-switches the cavity, or a second (possibly identical) modulator. A dumped cavity is more complicated to align than simple Q-switching, and may need a control loop to choose the best time at which to dump the beam from the cavity.

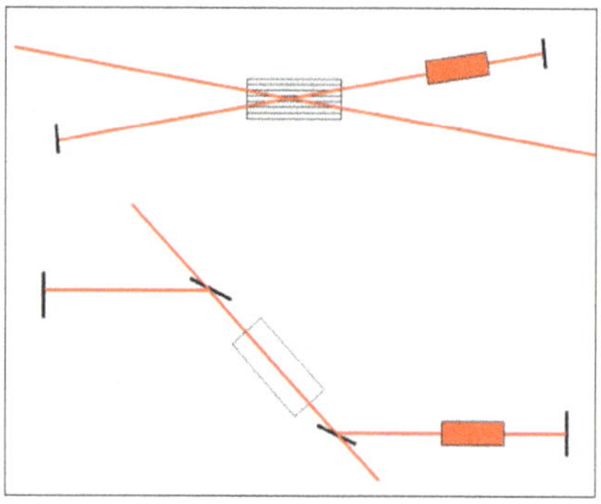

Regenerative amplifier. Red line: Laser beam. Red box: Gain medium. Top: AOM-based design. Bottom: The Pockel's cell-based design needs thin film polarizers. The direction of the emitted pulse depends on the timing.

In regenerative amplification, an optical amplifier is placed inside a Q-switched cavity. Pulses of light from another laser (the "master oscillator") are injected into the cavity by lowering the Q to allow the pulse to enter and then increasing the Q to confine the pulse to the cavity where it can be amplified by repeated passes through the gain medium. The pulse is then allowed to leave the cavity via another Q switch.

Typical Performance

A typical Q-switched laser (e.g. a Nd:YAG laser) with a resonator length of e.g. 10 cm can produce light pulses of several tens of nanoseconds duration. Even when the average power is well below 1 W, the peak power can be many kilowatts. Large-scale laser systems can produce Q-switched pulses with energies of many joules and peak powers in the gigawatt region. On the other hand, passively Q-switched microchip lasers (with very short resonators) have generated pulses with durations far below one nanosecond and pulse repetition rates from hundreds of hertz to several megahertz (MHz).

Applications

Q-switched lasers are often used in applications which demand high laser intensities in nanosecond pulses, such as metal cutting or pulsed holography. Nonlinear optics often takes advantage of the high peak powers of these lasers, offering applications such as 3D optical data storage and 3D microfabrication. However, Q-switched lasers can also be used for measurement purposes, such as for distance measurements (range finding) by measuring the time it takes for the pulse to get to some target and the reflected light to get back to the sender. It can be also used in chemical dynamic study, e.g. temperature jump relaxation study.

Q-switched lasers are also used to remove tattoos by shattering ink pigments into particles that are cleared by the body's lymphatic system. Full removal can take between six and twenty treatments depending on the amount and colour of ink, spaced at least a month apart, using different wavelengths for different coloured inks. Nd:YAG lasers are currently the most favoured lasers due to their high peak powers, high repetition rates and relatively low costs. In 2013 a picosecond laser was introduced based on clinical research which appears to show better clearance with 'difficult' colours such as green and light blue.

Q-switch laser is also used by beauticians around the world to treat skin-related issues like acne, pigmentation, dark spots, and fixes for anti-aging. According to *The Telegraph*, in India, leading skin care specialists like Kaya Skin Clinic, Richfeel, and VLCC make use of Q-switch laser technology for a variety of purposes. A review of these services provided to the customers at large denote that Q switch is a more effective alternative than most other techniques.

Laser Pumping

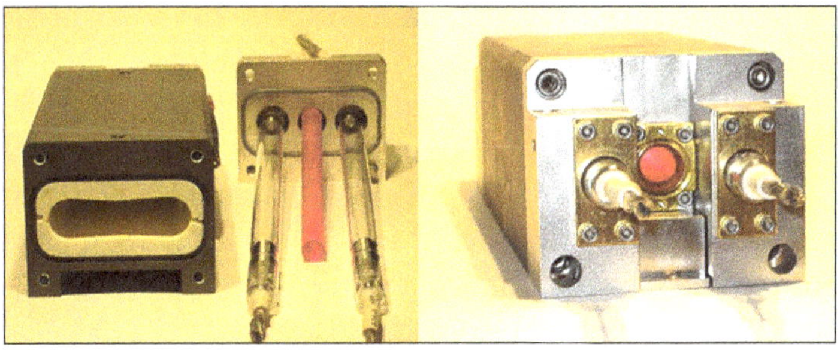

A ruby laser head. The photo on the left shows the head unassembled, revealing the pumping cavity, the rod and the flashlamps. The photo on the right shows the head assembled.

Laser pumping is the act of energy transfer from an external source into the gain medium of a laser. The energy is absorbed in the medium, producing excited states in its atoms. When the number of particles in one excited state exceeds the number of particles in the ground state or a less-excited state, population inversion is achieved. In this condition, the mechanism of stimulated emission can take place and the medium can act as a laser or an optical amplifier. The pump power must be higher than the lasing threshold of the laser.

The pump energy is usually provided in the form of light or electric current, but more exotic sources have been used, such as chemical or nuclear reactions.

Optical Pumping

Pumping Cavities

A laser pumped with an arc lamp or a flashlamp is usually pumped through the lateral wall of the lasing medium, which is often in the form of a crystal rod containing a metallic impurity or a glass tube containing a liquid dye, in a condition known as "side-pumping." To use the lamp's energy most efficiently, the lamps and lasing medium are contained in a reflective cavity that will redirect most of the lamp's energy into the rod or dye cell.

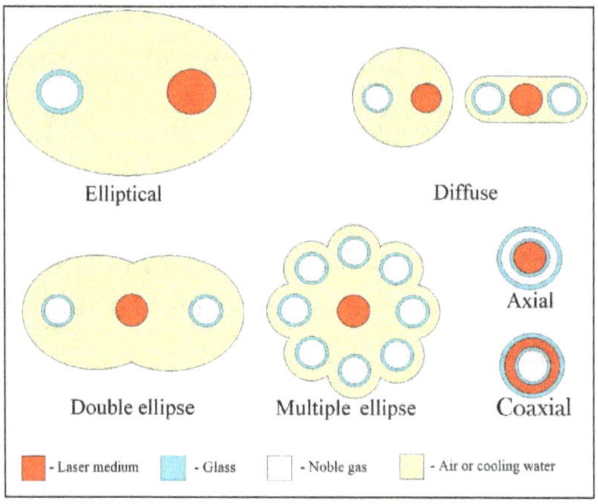

Various laser pumping cavity configurations.

In the most common configuration, the gain medium is in the form of a rod located at one focus of a mirrored cavity, consisting of an elliptical cross-section perpendicular to the rod's axis. The flashlamp is a tube located at the other focus of the ellipse. Often the mirror's coating is chosen to reflect wavelengths that are shorter than the lasing output while absorbing or transmitting wavelengths that are the same or longer, to minimize thermal lensing. In other cases an absorber for the longer wavelengths is used. Often, the lamp is surrounded by a cylindrical jacket called a flow tube. This flow tube is usually made of a glass that will absorb unsuitable wavelengths, such as ultraviolet, or provide a path for cooling water which absorbs infrared. Often, the jacket is given a dielectric coating that reflects unsuitable wavelengths of light back into the lamp. This light is absorbed and some of it is re-emitted at suitable wavelengths. The flow tube also serves to protect the rod in the event of a violent lamp failure.

Smaller ellipses create fewer reflections, (a condition called "close-coupling"), giving higher intensity in the center of the rod. For a single flashlamp, if the lamp and rod are equal diameter, an ellipse that is twice as wide as it is high is usually the most efficient at imaging the light into the rod. The rod and the lamp are relatively long to minimize the effect of losses at the end faces and to provide a sufficient length of gain medium. Longer flashlamps are also more efficient at transferring electrical energy into light, due to higher impedance. However, if the rod is too long in relation to its diameter a condition called "prelasing" can occur, depleting the rod's energy before it can properly build up. Rod ends are often antireflection coated or cut at Brewster's angle to minimize this effect. Flat mirrors are also often used at the ends of the pump cavity to reduce loss.

Variations on this design use more complex mirrors composed of overlapping elliptical shapes, to allow multiple flashlamps to pump a single rod. This allows greater power, but are less efficient because not all of the light is correctly imaged into the rod, leading to increased thermal losses. These losses can be minimized by using a close-coupled cavity. This approach may allow more symmetric pumping, increasing beam quality, however.

Another configuration uses a rod and a flashlamp in a cavity made of a diffuse reflecting material, such as spectralon or powdered barium sulfate. These cavities are often circular or oblong, as focusing the light is not a primary objective. This doesn't couple the light as well into the lasing medium, since the light makes many reflections before reaching the rod, but often requires less maintenance than metalized reflectors. The increased number of reflections is compensated for by the diffuse medium's higher reflectivity: 99% compared to 97% for a gold mirror. This approach is more compatible with unpolished rods or multiple lamps.

Parasitic modes occur when reflections are generated in directions other than along the length of the rod, which can use up energy that would otherwise be available to the beam. This can be a particular problem if the barrel of the rod is polished. Cylindrical laser rods support whispering gallery modes due to total internal reflection between the rod and the cooling water, which reflect continuously around the circumference of the rod. Light pipe modes can reflect down the length of the rod in a zig-zag path. If the rod has an antireflection coating, or is immersed in a fluid that matches its refractive index, it can dramatically reduce these parasitic reflections. Likewise, if the barrel of the rod is rough ground (frosted), or grooved, internal reflections can be dispersed.

Pumping with a single lamp tends to focus most of the energy on one side, worsening the beam profile. It is common for rods to have a frosted barrel, to diffuse the light, providing a more even distribution of light throughout the rod. This allows more energy absorption throughout the gain medium for a better transverse mode. A frosted flow tube or diffuse reflector, while leading to lowered transfer efficiency, helps increase this effect, improving the gain.

Laser host materials are chosen to have a low absorption; only the dopant absorbs. Therefore, any light at frequencies not absorbed by the doping will go back into the lamp and reheat the plasma, shortening lamp life.

Flashlamp Pumping

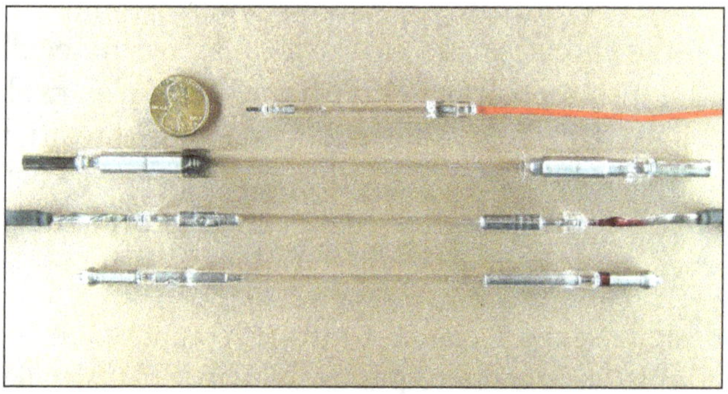

Laser pumping lamps. The top three are xenon flashlamps while the bottom one is a krypton arc lamp.

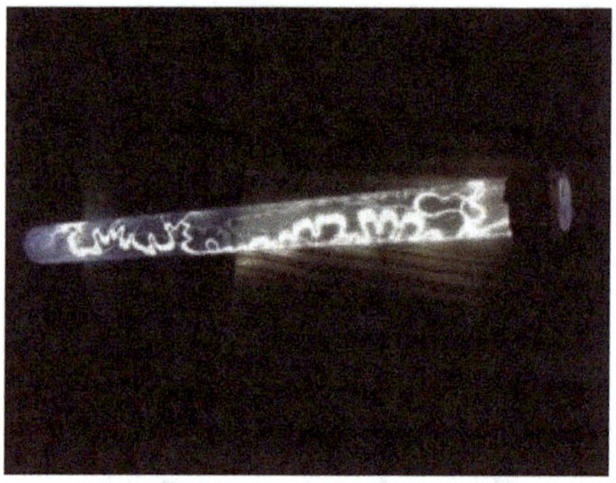

External triggering was used in this extremely fast discharge. Due to the very high speed, (3.5 microseconds), the current is not only unable to fully heat the xenon and fill the tube, but is still in direct contact with the glass.

Flashlamps were the earliest energy source for lasers. They are used for high pulsed energies in both solid-state and dye lasers. They produce a broad spectrum of light, causing most of the energy to be wasted as heat in the gain medium. Flashlamps also tend to have a short lifetime. The first laser consisted of a helical flashlamp surrounding a ruby rod.

Quartz flashlamps are the most common type used in lasers, and, at low energies or high repetition rates, can operate at temperatures as high as 900 °C. Higher average powers or repetition rates require water cooling. The water usually has to wash across not only the arc length of the lamp, but across the electrode portion of the glass as well. Water-cooled flashlamps are usually manufactured with the glass shrunken around the electrode to allow direct cooling of the tungsten. If the electrode is allowed to heat much more than the glass thermal expansion can crack the seal.

Lamp lifetime depends primarily on the energy regime used for the particular lamp. Low energies give rise to sputter, which can remove material from the cathode and redeposit it on the glass, creating a darkened, mirrored appearance. The life expectancy at low energies can be quite unpredictable. High energies cause wall ablation, which not only gives the glass a cloudy appearance, but also weakens it structurally and releases oxygen, affecting pressure, but at these energy levels the life expectancy can be calculated with a fair amount of accuracy.

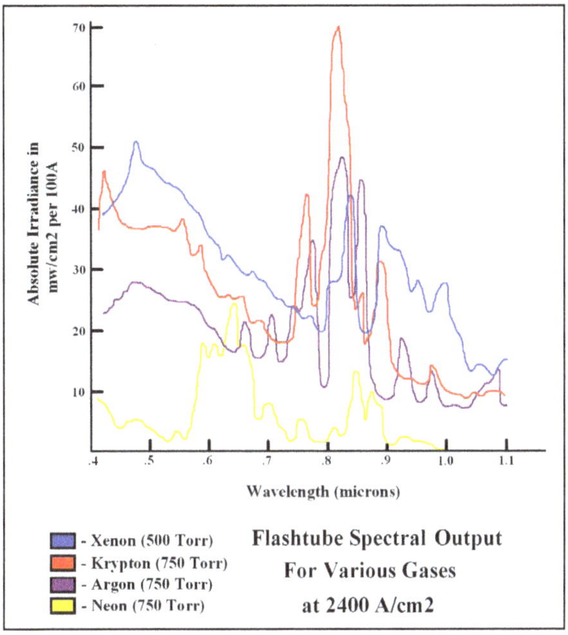

The spectral outputs for flashlamps using various gases, at a current density approaching that of greybody radiation.

Pulse duration can also affect lifetime. Very long pulses can strip large amounts of material from the cathode, depositing it on the walls. With very short pulse durations, care must be taken to ensure that the arc is centered in the lamp, far away from the glass, preventing serious wall ablation. External triggering is not usually recommended for short pulses. Simmer voltage triggering is usually used for extremely fast discharges, as are used in dye lasers, and often combine this with a "pre-pulse technique", where as a small flash is initiated just milliseconds before the main flash, to preheat the gas for a faster rise time.

Dye lasers sometimes use "axial pumping," which consists of a hollow, annular shaped flashlamp, with the outer envelope mirrored to reflect suitable light back to the center. The dye cell is placed in the middle, providing a more even distribution of pumping light, and more efficient transfer of energy. The hollow flashlamp also has lower inductance than a normal flashlamp, which provides a shorter flash discharge. Rarely, a "co-axial" design is used for dye lasers, which consists of a normal flashlamp surrounded by an annular shaped dye cell. This provides better transfer efficiency, eliminating the need for a reflector, but diffraction losses cause a lower gain.

The output spectrum of a flashlamp is primarily a product of its current density. After determining the "explosion energy" for the pulse duration, (the amount of energy that will destroy it in one to ten flashes), and choosing a safe energy level for operation, the balance of voltage and capacitance can be adjusted to center the output anywhere from the near infrared to the far ultraviolet. Low current densities result from the use of very high voltage and low current. This produces broadened spectral lines with the output centered in the near-IR, and is best for pumping infrared lasers such as Nd:YAG and erbium:YAG. Higher current densities broaden the spectral lines to the point where they begin to blend together, and continuum emission is produced. Longer wavelengths reach saturation levels at lower current densities than shorter wavelengths, so as current is increased the output center will shift toward the visual spectrum, which is better for pumping visible light lasers, such as ruby. At this point, the gas becomes nearly an ideal "greybody radiator." Even higher current densities will produce blackbody radiation, centering the output in the ultraviolet.

Xenon is used extensively because of its good efficiency, although krypton is often used for pumping neodymium doped laser rods. This is because the spectral lines in the near-IR range better match the absorption lines of neodymium, giving krypton better transfer efficiency even though its overall power output is lower. This is especially effective with Nd:YAG, which has a narrow absorption profile. Pumped with krypton, these lasers can achieve up to twice the output power obtainable from xenon. Spectral line emission is usually chosen when pumping Nd:YAG with krypton, but since all of xenon's spectral lines miss the absorption bands of Nd:YAG, when pumping with xenon the continuum emission is used.

Arc Lamp Pumping

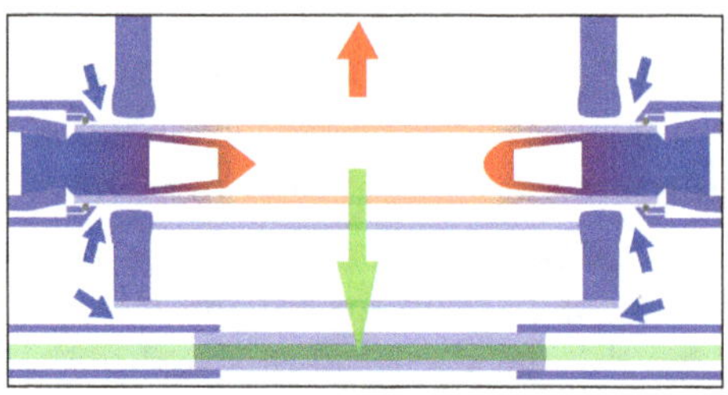

Optical pumping of a laser rod (bottom) with an arc lamp (top). Red: hot. Blue: cold. Green: light. Non-green arrows: water flow. Solid colors: metal. Light colors: fused quartz.

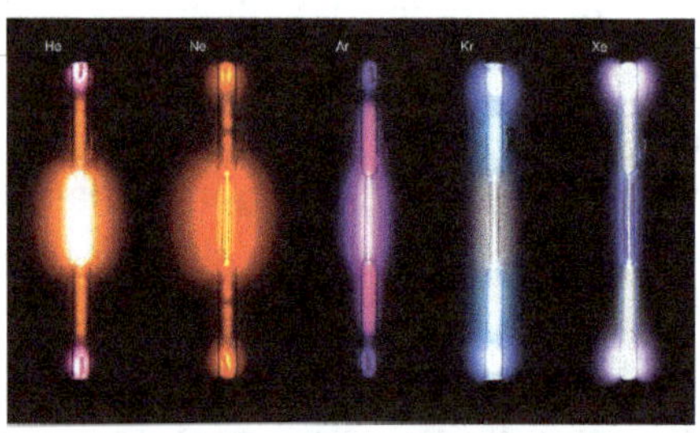

These gas-discharge lamps show the spectral line outputs of the various noble gases.

Arc lamps are used for pumping rods that can support continuous operation, and can be made any size and power. Typical arc lamps operate at a voltage high enough to maintain the certain current level for which the lamp was designed to operate. This is often in the range of 10 to 50 amps. Due to their very high pressures, arc lamps require specially designed circuitry for start up, or "striking" the arc. Striking usually occurs in three phases. In the triggering phase, an extremely high voltage pulse from the "series triggering" transformer creates a spark streamer between the electrodes, but the impedance is too high for the main voltage to take over. A "boost voltage" phase is then initiated, where a voltage that is higher than the voltage drop between the electrodes is driven through the lamp, until the gas is heated to a plasma state. When impedance becomes low enough, the "current control" phase takes over, where as the main voltage begins to drive the current to a stable level.

Arc lamp pumping takes place in a cavity similar to a flashlamp pumped laser, with a rod and one or more lamps in a reflector cavity. The exact shape of the cavity is often dependent on how many lamps are used. The main difference is in the cooling. Arc lamps need to be cooled with water, ensuring that the water washes beyond the glass, and across the electrode connectors as well. This requires the use of deionized water with a resistivity of at least 200 kilohms, to keep from shorting out the circuit and corroding the electrodes through electrolysis. Water is typically channeled through a flow tube at a rate of 4 to 10 liters per minute.

Arc lamps come in nearly all of the noble gas types, including xenon, krypton, argon, neon, and helium, which all emit spectral lines that are very specific to the gas. The output spectrum of an arc lamp is mostly dependent on the gas type, being narrow band spectral lines very similar to a flashlamp operated at low current densities. The output is highest in the near infrared, and are usually used to pump infrared lasers such as Nd:YAG.

External Laser Pumping

A dye laser tuned to 589nm (amber yellow), pumped with an external, frequency-doubled Nd:YAG laser @ 532nm (yellowish-green). The closeness between wavelengths results in a very small Stokes shift, reducing energy losses.

A laser of a suitable type can be used to pump another laser. The pump laser's narrow spectrum allows it to be closely matched to the absorption lines of the lasing media, giving it much more efficient energy transfer than the broadband emission of flashlamps. Diode lasers pump solid state lasers and liquid dye lasers. A ring laser design is often used, especially in dye lasers. The ring laser uses three or more mirrors to reflect light in a circular path. This helps eliminate the standing wave generated by most Fabry–Pérot resonators, leading to a better use of the gain medium's energy.

Other Optical Pumping Methods

Microwaves or radiofrequency EM radiation can be used to excite gas lasers. A solar-pumped laser uses solar radiation as a pump source.

Electrical Pumping

Electric glow discharge is common in gas lasers. For example, in the helium–neon laser the electrons from the discharge collide with the helium atoms, exciting them. The excited helium atoms then collide with neon atoms, transferring energy. This allows an inverse population of neon atoms to build up.

Electric current is typically used to pump laser diodes and semiconductor crystal lasers (for example germanium). Electron beams pump free electron lasers and some excimer lasers.

Gas Dynamic Pumping

Gas dynamic lasers are constructed using the supersonic flow of gases, such as carbon dioxide, to excite the molecules past threshold. The gas is pressurized and then heated to as high as 1400 kelvins. The gas is then allowed to expand rapidly through specially shaped nozzles to a very low pressure. This expansion occurs at supersonic velocities, sometimes as high as mach 4. The hot gas has many molecules in the upper excited states, while many more are in the lower states. The rapid expansion causes adiabatic cooling, which reduces the temperature to as low as 300 K. This reduction in temperature causes the molecules in the upper and lower states to relax their equilibrium to a value that is more appropriate for the lower temperature. However, the molecules in the lower states relax very quickly, while the upper state molecules take much longer to relax. Since a good quantity of molecules remain in the upper state, a population inversion is created, which often extends for quite a distance downstream. Continuous wave outputs as high as 100 kilowatts have been obtained from dynamic carbon dioxide lasers.

Similar methods of supersonic expansion are used to adiabatically cool carbon monoxide lasers, which are then pumped either through chemical reaction, electrical, or radio frequency pumping. The adiabatic cooling replaces bulky and costly cryogenic cooling with liquid nitrogen, increasing the carbon monoxide laser's efficiency. Lasers of this type have been able to produce outputs as high as a gigawatt, with efficiencies as high as 60%.

Other Types

Charge-displacement self-channeling can give rise to high energy concentration along a column created and maintained by the ponderomotive expulsion of electrons. The channel will also columnate shorter wavelength secondary radiation and ultimately extremely short wavelength lasing.

Chemical reaction is used as a power source in chemical lasers. This allows for very high output powers difficult to reach by other means.

Nuclear fission is used in exotic nuclear pumped lasers (NPL), directly employing the energy of the fast neutrons released in a nuclear reactor.

The United States military tested an X-ray laser pumped by a nuclear weapon in the 1980s, but the results of the test were inconclusive and it has not been repeated.

Active Laser Medium

The active laser medium (also called gain medium or lasing medium) is the source of optical gain within a laser. The gain results from the stimulated emission of electronic or molecular transitions to a lower energy state from a higher energy state previously populated by a pump source.

Examples of active laser media include:

- Certain crystals, typically doped with rare-earth ions (e.g. neodymium, ytterbium, or erbium) or transition metal ions (titanium or chromium); most often yttrium aluminium garnet ($Y_3Al_5O_{12}$), yttrium orthovanadate (YVO_4), or sapphire (Al_2O_3); and not often Caesium cadmium bromide ($CsCdBr_3$).
- Glasses, e.g. silicate or phosphate glasses, doped with laser-active ions.
- Gases, e.g. mixtures of helium and neon (HeNe), nitrogen, argon, carbon monoxide, carbon dioxide, or metal vapors.
- Semiconductors, e.g. gallium arsenide (GaAs), indium gallium arsenide (InGaAs), or gallium nitride (GaN).
- Liquids, in the form of dye solutions as used in dye lasers.

In order to fire a laser, the active gain medium must be in a nonthermal energy distribution known as a population inversion. The preparation of this state requires an external energy source and is known as laser pumping. Pumping may be achieved with electrical currents (e.g. semiconductors, or gases via high-voltage discharges) or with light, generated by discharge lamps or by other lasers (semiconductor lasers). More exotic gain media can be pumped by chemical reactions, nuclear fission, or with high-energy electron beams.

Example of a model of gain medium:

A universal model valid for all laser types does not exist. The simplest model includes two systems of sub-levels: upper and lower. Within each sub-level system, the fast transitions ensure that thermal equilibrium is reached quickly, leading to the Maxwell–Boltzmann statistics of excitations among sub-levels in each system. The upper level is assumed to be metastable. Also, gain and refractive index are assumed independent of a particular way of excitation.

For good performance of the gain medium, the separation between sub-levels should be larger than working temperature; then, at pump frequency ω_P the absorption dominates.

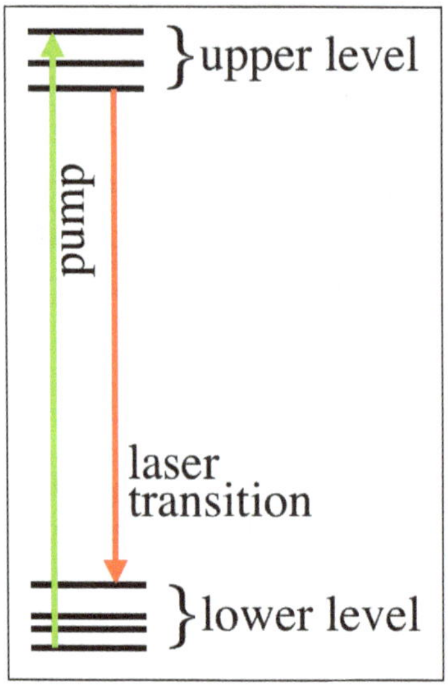

Simplified scheme of levels a gain medium.

In the case of amplification of optical signals, the lasing frequency is called *signal frequency*. However, the same term is used even in the laser oscillators, when amplified radiation is used to transfer energy rather than information. The model below seems to work well for most optically-pumped solid-state lasers.

Cross-sections

The simple medium can be characterized with effective cross-sections of absorption and emission at frequencies ω_P and ω_S.

- Have N be concentration of active centers in the solid-state lasers.
- Have N_1 be concentration of active centers in the ground state.
- Have N_2 be concentration of excited centers.
- Have $N_1 + N_2 = N$.

The relative concentrations can be defined as $n_1 = N_1 / N$ and $n_2 = N_2 / N$.

The rate of transitions of an active center from ground state to the excited state can be expressed with $W_{\mathrm{u}} = \dfrac{I_{\mathrm{p}}\sigma_{\mathrm{ap}}}{\hbar\omega_{\mathrm{p}}} + \dfrac{I_{\mathrm{s}}\sigma_{\mathrm{as}}}{\hbar\omega_{\mathrm{s}}}$.

The rate of transitions back to the ground state can be expressed with $W_d = \frac{I_p \sigma_{ep}}{\hbar\omega_p} + \frac{I_s \sigma_{es}}{\hbar\omega_s} + \frac{1}{\tau}$ where, σ_{as} and σ_{ap} are effective cross-sections of absorption at the frequencies of the signal and the pump.σ_{es} and σ_{ep} are the same for stimulated emission. $\frac{1}{\tau}$ is rate of the spontaneous decay of the upper level.

Then, the kinetic equation for relative populations can be written as follows:

$$\frac{dn_2}{dt} = W_u n_1 - W_d n_2$$

$$\frac{dn_1}{dt} = -W_u n_1 + W_d n_2$$

However, these equations keep $n_1 + n_2 = 1$.

The absorption A at the pump frequency and the gain G at the signal frequency can be written as follows:

$$A = N_1 \sigma_{pa} - N_2 \sigma_{pe}$$

$$G = N_2 \sigma_{se} - N_1 \sigma_{sa} \cdot$$

Steady-state Solution

In many cases the gain medium works in a continuous-wave or quasi-continuous regime, causing the time derivatives of populations to be negligible.

The steady-state solution can be written:

$$n_2 = \frac{W_u}{W_u + W_d}, \quad n_1 = \frac{W_d}{W_u + W_d}.$$

The dynamic saturation intensities can be defined:

$$I_{po} = \frac{\hbar\omega_p}{(\sigma_{ap} + \sigma_{ep})\tau}, \quad I_{so} = \frac{\hbar\omega_s}{(\sigma_{as} + \sigma_{es})\tau}.$$

The absorption at strong signal: $A_0 = \frac{ND}{\sigma_{as} + \sigma_{es}}$.

The gain at strong pump: $G_0 = \frac{ND}{\sigma_{ap} + \sigma_{ep}}$

where, $D = \sigma_{pa}\sigma_{se} - \sigma_{pe}\sigma_{sa}$ is determinant of cross-section.

Gain never exceeds value G_0 and absorption never exceeds value A_0U.

At given intensities I_p, I_s of pump and signal, the gain and absorption can be expressed as follows:

$$A = A_0 \frac{U+s}{1+p+s}, \quad G = G_0 \frac{p-V}{1+p+s},$$

Where, $p = I_p / I_{po}$, $s = I_s / I_{so}$, $U = \frac{(\sigma_{as}+\sigma_{es})\sigma_{ap}}{D}$, $V = \frac{(\sigma_{ap}+\sigma_{ep})\sigma_{as}}{D}$.

Identities

The following identities take place:

$$U - V = 1, A / A_0 + G / G_0 = 1.$$

The state of gain medium can be characterized with a single parameter, such as population of the upper level, gain or absorption.

Efficiency of the Gain Medium

The efficiency of a gain medium can be defined $E = \frac{I_s G}{I_p A}$.

Within the same model, the efficiency can be expressed as follows:

$$E = \frac{\omega_s}{\omega_p} \frac{1-V/p}{1+U/s}.$$

For the efficient operation both intensities, pump and signal should exceed their saturation intensities:

$$\frac{p}{V} \gg 1, and \frac{s}{U} \gg 1.$$

The estimates above are valid for a medium uniformly filled with pump and signal light. Spatial hole burning may slightly reduce the efficiency because some regions are pumped well, but the pump is not efficiently withdrawn by the signal in the nodes of the interference of counter-propagating waves.

References

- Früngel, Frank B. A. (2014). Optical Pulses - Lasers - Measuring Techniques. Academic Press. P. 192. ISBN 9781483274317. Retrieved 1 February 2015
- Laser-physics-204, definitions, homework-help: chegg.com, Retrieved 11 March, 2019

- Reiner, J. E.; Robertson, J. W. F.; Burden, D. L.; Burden, L. K.; Balijepalli, A.; Kasianowicz, J. J. (2013). "Temperature Sculpting in Yoctoliter Volumes". Journal of the American Chemical Society. 135: 3087–3094. doi:10.1021/ja309892e. ISSN 0002-7863. PMC 3892765. PMID 23347384
- Gain-switching: rp-photonics.com, Retrieved 12 April, 2015
- "Lamp-pumped Lasers". Encyclopedia of Laser Physics and Technology. RP Photonics. Retrieved 3 February 2009
- Reiner, J. E.; Robertson, J. W. F.; Burden, D. L.; Burden, L. K.; Balijepalli, A.; Kasianowicz, J. J. (2013). "Temperature Sculpting in Yoctoliter Volumes". Journal of the American Chemical Society. 135: 3087–3094. Doi:10.1021/ja309892e. ISSN 0002-7863. PMC 3892765. PMID 23347384

Gas Laser and its Types

3

• Types of Gas Lasers

A laser in which an electric current is discharged through a gas for producing coherent light is referred to as gas laser. Various types of gas laser include helium–neon lasers, chemical lasers, excimer lasers, ion lasers and metal-vapor lasers. The significant aspects of these types of gas lasers have been thoroughly discussed in this chapter.

A gas laser is a laser with a gaseous active medium. The laser-active entities are either single atoms or molecules, and are often used in a mixture with other substances having auxiliary functions. A population inversion as the prerequisite for gain via stimulated emission is in most cases achieved by pumping the gas with an electric discharge, but there are also gas lasers using a chemical reaction, optically pumped devices, and Raman lasers. During operation, the gas is often in the state of a plasma, containing a significant concentration of electrically charged particles.

Most gas lasers emit with a high beam quality, often close to diffraction-limited, since the gas introduces only weak optical distortions, despite considerable temperature gradients. Their operation usually requires a high-voltage supply, often with a high electrical power. Some high-power gas lasers use a system for quickly circulating the gas (*forced convection, fast flow*).

There are very different kinds of gas lasers, operating in entirely different regimes concerning emission wavelength and output power:

- Helium–neon lasers (He–Ne lasers) often emit red light at 632.8 nm, but can also be made for other wavelengths such as 543.5 nm (green), 594.1 nm (yellow), 611.9 nm (orange), 3.39 μm, or 1.15 μm. Typical He–Ne lasers have a gas cell with a length of the order of 20 cm and generate a few milliwatts of output power in continuous-wave operation at 632.8 nm, using several watts of electrical power. Helium–neon lasers are often used for alignment and in interferometers, and compete with laser diodes, which are more compact and efficient. Some HeNe lasers serve in optical frequency standards.
- Argon ion lasers use a typically larger (e.g. 1 m long) water-cooled tube with an argon plasma, made with an electrical discharge with high current density in order to achieve a high degree of ionization. They can generate more than 20 W

of output power in green light at 514.5 nm, and less at some other wavelengths such as 457.9, 488.0, or 351 nm. Their power efficiency is fairly low, so that tens of kilowatts of electrical power are required for multi-watt green output, and the cooling system has corresponding dimensions. There are smaller tubes for air-cooled argon lasers, requiring hundreds of watts for generating some tens of milliwatts. Argon ion lasers can be used e.g. for pumping titanium–sapphire lasers and dye lasers, and are rivaled by frequency-doubled diode-pumped solid-state lasers.

- Krypton ion lasers are similar to argon ion lasers and can emit high powers at 647.1 nm and some other wavelengths.
- Carbon dioxide lasers (CO_2 lasers) use a gas mixture of CO_2, helium (He), nitrogen (N_2), and possibly some hydrogen (H_2), water vapor, and/or xenon (Xe) for generating laser radiation mostly at 10.6 μm. They have wall-plug efficiencies above 10% and are suitable for output powers of multiple kilowatts with fairly high beam quality. They are widely used for material processing, e.g. cutting, welding and marking, but also in laser surgery. CO_2 lasers are called molecular lasers because laser-active molecules are used.
- Carbon monoxide lasers (CO lasers) can have wall-plug efficiencies of the order of 40%, thus being substantially more power-efficient than CO_2 lasers. They can emit on various lines between 4.8 μm and 8.3 μm and are mostly used as light sources for laser absorption spectroscopy. Following technological advances concerning the device lifetime, CO lasers emitting around 5.5 μm might also become interesting for laser material processing (e.g. cutting of glasses); in comparison with the widespread CO_2 lasers, they offer better absorption in many materials and better focusing capabilities.
- Excimer lasers (rare gas halide lasers, exciplex lasers) are also pumped with an electrical discharge, but in that case the pumping energy is used to form unstable molecules which can emit photons when disassociating. Most excimer lasers are ultraviolet lasers and are operated with current pulses, leading to the emission of intense nanosecond pulses. They are used for various types of material processing, including pulsed laser deposition, laser marking, and the fabrication of fiber Bragg gratings. There are also medical applications e.g. in ophthalmology.
- Nitrogen lasers are another type of pulsed ultraviolet laser, based on pure nitrogen, a nitrogen–helium mixture, and sometimes even simply air (with lower performance). Emission typically occurs at 337.1 nm in the form of short pulses; a self-terminating laser transition is used. The high gain leads to relatively efficient superluminescent emission even without a laser resonator. Nitrogen lasers are relatively easy to build and operate, and have been made by many hobbyists without refined laboratory equipment.

- Hydrogen lasers can be used to access even shorter wavelengths around 160 nm, 123 nm or 116 nm.
- Various metal vapor lasers use a metal vapor, excited and heated by an electric discharge.

Copper vapor lasers are excited with intense current pulses and generate nanosecond pulses at 510.6 nm (green) or 578.2 nm (yellow). The average output power can exceed 100 W. *Helium–cadmium lasers* are more similar to helium–neon lasers, emitting continuously at 442 nm (blue) or 325 nm (ultraviolet), with optical powers of the order of 100 mW. The laser transition occurs in Cd^+ ions, which become excited in collisions with excited helium atoms. - There are *alkali vapor lasers*, using e.g. a cesium or rubidium cell in an oven as the gain medium. Such lasers can be pumped with laser diodes. The power efficiency can be fairly high; note that the quantum defect can be small for a typical pumping scheme where one pumps from the ground state $^6S_{1/2}$ to $^6P_{3/2}$ and uses a transfer via a buffer gas (e.g. ethane) to the nearby $^6P_{1/2}$ as the upper laser level. At the same time, the beam quality can be much higher than that of the pump source, so that such a laser acts as an efficient brightness converter. - *Chemical lasers* convert chemical energy of gases into laser light (typically in the mid- or near-infrared region) with powers up to the megawatt level. There are e.g. hydrogen-fluoride (HF) lasers, fueled with H_2 and F_2, which is converted to HF, and oxygen-iodine lasers (COIL). Chemical lasers are mainly studied for military purposes, e.g. as anti-missile weapons, to be operated even on board large airplanes. -Raman gas lasers are Raman lasers, based on optical amplification via stimulated Raman scattering rather than on stimulated emission of excited ions. They can e.g. use a hydrogen cell, and need to be optically pumped.

Gas lasers can also be grouped according to the nature of their laser-active species:

- Neutral atom gas lasers include helium–neon lasers and copper vapor lasers.
- Ion lasers use free ions; examples are helium–cadmium lasers, argon ion lasers and krypton lasers. Typically, ion lasers generate shorter wavelengths, but with moderate power efficiency.
- Molecular gas lasers use gas molecules. Examples are carbon dioxide and carbon monoxide lasers, nitrogen lasers, and excimer lasers.

Many gas lasers have self-terminating laser transitions, where the lower state has a long lifetime. Lasing stops once the lower-state population becomes too high. Examples for such gas lasers are nitrogen lasers and copper vapor lasers. Excimer lasers can also only be operated in pulsed mode, although for different reasons.

Application of Gas Lasers

A primary reason for using certain gas lasers instead of solid-state lasers is that they offer special wavelengths, which are otherwise difficult to obtain. Another interesting aspect is that relatively high optical powers can be obtained with gas lasers; compared to solid-state lasers, particularly diode-pumped ones, the price depends less strongly on the required output power level. A good example for both aspects is the carbon dioxide laser, which is quite a unique long-wavelength source with high output power. Similarly, excimer lasers provide high powers in the ultraviolet spectral region.

The helium–neon laser has been widely used for the generation of red laser light, but is now increasingly replaced by cheaper and more compact laser diodes. Similarly, argon ion lasers were often used for pumping titanium–sapphire lasers, but are now often replaced by frequency-doubled solid-state lasers.

Types of Gas Lasers

Helium–Neon Laser

A helium–neon laser or HeNe laser, is a type of gas laser whose gain medium consists of a mixture of 90% helium and 10% neon at a total pressure of about 1 mm of Hg inside of a small electrical discharge. The best-known and most widely used HeNe laser operates at a wavelength of 632.8 nm, in the red part of the visible spectrum.

The first HeNe lasers emitted infrared at 1.15 μm, and were the first gas lasers. However, a laser that operated at visible wavelengths was much more in demand, and a number of other neon transitions were investigated to identify ones in which a population inversion can be achieved. The 633 nm line was found to have the highest gain in the visible spectrum, making this the wavelength of choice for most HeNe lasers. However other visible as well as infrared stimulated emission wavelengths are possible, and by using mirror coatings with their peak reflectance at these other wavelengths, HeNe lasers could be engineered to employ those transitions; this includes visible lasers appearing red, orange, yellow, and green. Stimulated emissions are known from over 100 μm in the far infrared to 540 nm in the visible. Because visible transitions have somewhat lower gain, these lasers generally have lower output efficiencies and are more costly. The 3.39 μm transition has a very high gain but is prevented from use in an ordinary HeNe laser (of a different intended wavelength) because the cavity and mirrors are lossy at that wavelength. However in high power HeNe lasers having a particularly long cavity, superluminescence at 3.39 μm can become a nuisance, robbing power from the stimulated emission medium, often requiring additional suppression. The best-known and most widely used HeNe laser operates at a wavelength of 632.8 nm, in the red part of the visible spectrum. It was developed at Bell Telephone Laboratories in 1962, 18 months after the pioneering demonstration at the same laboratory of the first continuous infrared HeNe gas laser in December 1960.

Construction and Operation

The gain medium of the laser, as suggested by its name, is a mixture of helium and neon gases, in approximately a 10:1 ratio, contained at low pressure in a glass envelope. The gas mixture is mostly helium, so that helium atoms can be excited. The excited helium atoms collide with neon atoms, exciting some of them to the state that radiates 632.8 nm. Without helium, the neon atoms would be excited mostly to lower excited states responsible for non-laser lines. A neon laser with no helium can be constructed but it is much more difficult without this means of energy coupling. Therefore, a HeNe laser that has lost enough of its helium (e.g., due to diffusion through the seals or glass) will lose its laser functionality because the pumping efficiency will be too low. The energy or pump source of the laser is provided by a high voltage electrical discharge passed through the gas between electrodes (anode and cathode) within the tube. A DC current of 3 to 20 mA is typically required for CW operation. The optical cavity of the laser usually consists of two concave mirrors or one plane and one concave mirror, one having very high (typically 99.9%) reflectance and the output coupler mirror allowing approximately 1% transmission.

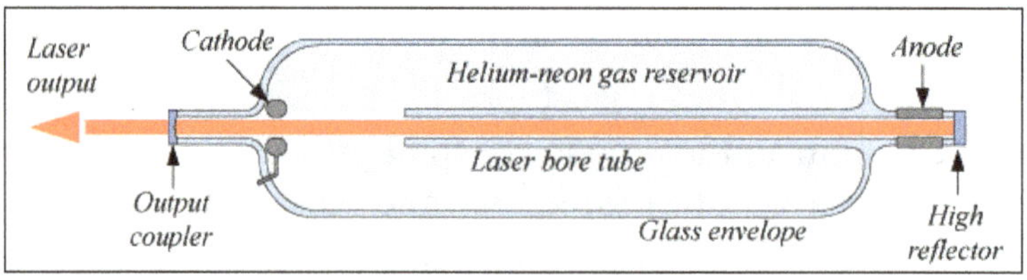

Schematic diagram of a helium–neon laser.

Commercial HeNe lasers are relatively small devices, among gas lasers, having cavity lengths usually ranging from 15 cm to 50 cm (but sometimes up to about 1 meter to achieve the highest powers), and optical output power levels ranging from 0.5 to 50 mW.

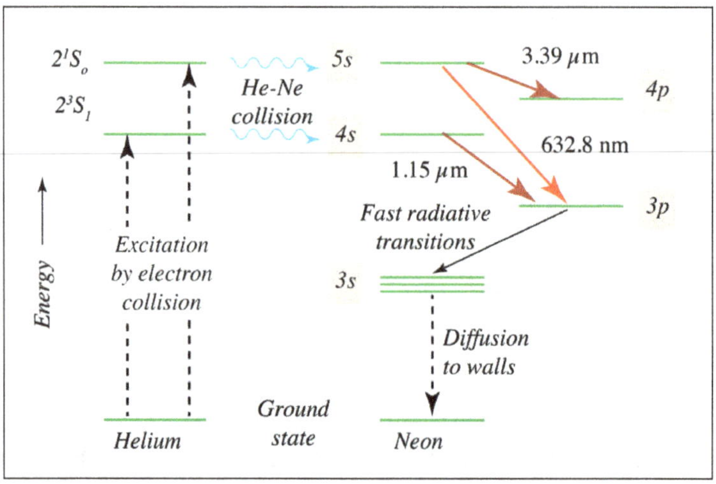

Energy levels in a He-Ne Laser.

The red HeNe laser wavelength of 633 nm has an actual vacuum wavelength of 632.991 nm, or about 632.816 nm in air. The wavelengths of the stimulated emission modes lie within about 0.001 nm above or below this value, and the wavelengths of those modes shift within this range due to thermal expansion and contraction of the cavity. Frequency-stabilized versions enable the wavelength of a single mode to be specified to within 1 part in 10^8 by the technique of comparing the powers of two longitudinal modes in opposite polarizations. Absolute stabilization of the laser's frequency (or wavelength) as fine as 2.5 parts in 10^{11} can be obtained through use of an iodine absorption cell.

The mechanism producing population inversion and light amplification in a HeNe laser plasma originates with inelastic collision of energetic electrons with ground state helium atoms in the gas mixture. As shown in the accompanying energy level diagram, these collisions excite helium atoms from the ground state to higher energy excited states, among them the 2^3S_1 and 2^1S_0 (LS or Russell-Saunder coupling, front number 2 tells that an excited electron is n = 2 state) in long-lived metastable states. Because of a fortuitous near coincidence between the energy levels of the two He metastable states, and the $5s_2$ and $4s_2$ (Paschen notation) levels of neon, collisions between these helium metastable atoms and ground state neon atoms results in a selective and efficient transfer of excitation energy from the helium to neon. This excitation energy transfer process is given by the reaction equations:

$$He^*(2^3S_1) + Ne^1S_0 \rightarrow He(^1S_0) + Ne^*4s_2 + \Delta E \text{ and}$$

$$He^*(2^1S) + Ne^1S_0 + \Delta E \rightarrow He(^1S_0) + Ne^*5s_2$$

Where (*) represents an excited state, and ΔE is the small energy difference between the energy states of the two atoms, of the order of 0.05 eV or 387 cm^{-1}, which is supplied by kinetic energy. Excitation energy transfer increases the population of the neon $4s_2$ and $5s_2$ levels manyfold. When the population of these two upper levels exceeds that of the corresponding lower level neon state, $3p_4$ to which they are optically connected, population inversion is present. The medium becomes capable of amplifying light in a narrow band at 1.15 μm (corresponding to the $4s_2$ to $3p_4$ transition) and in a narrow band at 632.8 nm (corresponding to the $5s_2$ to $3p_4$ transition at 632.8 nm). The $3p_4$ level is efficiently emptied by fast radiative decay to the 3s state, eventually reaching the ground state.

The remaining step in utilizing optical amplification to create an optical oscillator is to place highly reflecting mirrors at each end of the amplifying medium so that a wave in a particular spatial mode will reflect back upon itself, gaining more power in each pass than is lost due to transmission through the mirrors and diffraction. When these conditions are met for one or more longitudinal modes then radiation in those modes will rapidly build up until gain saturation occurs, resulting in a stable continuous laser beam output through the front (typically 99% reflecting) mirror.

The gain bandwidth of the HeNe laser is dominated by Doppler broadening rather than pressure broadening due to the low gas pressure, and is thus quite narrow: only about 1.5 GHz full width for the 633 nm transition. With cavities having typical lengths of 15 cm to 50 cm, this allows about 2 to 8 longitudinal modes to oscillate simultaneously (however single longitudinal mode units are available for special applications). The visible output of the red HeNe laser, long coherence length, and its excellent spatial quality, makes this laser a useful source for holography and as a wavelength reference for spectroscopy. A stabilized HeNe laser is also one of the benchmark systems for the definition of the meter.

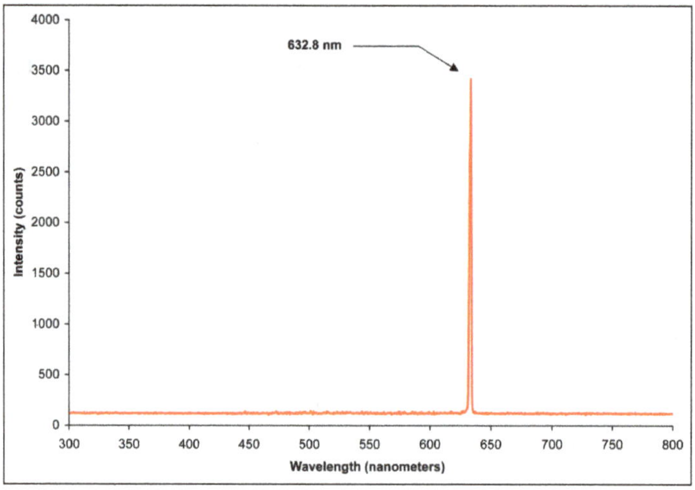

Spectrum of a helium neon laser illustrating its very high spectral purity (limited by the measuring apparatus). The 0.002 nm bandwidth of the stimulated emission medium is well over 10,000 times narrower than the spectral width of a light-emitting diode (whose spectrum is shown here for comparison), with the bandwidth of a single longitudinal mode being much narrower still.

Prior to the invention of cheap, abundant diode lasers, red HeNe lasers were widely used in barcode scanners at supermarket checkout counters. Laser gyroscopes have employed HeNe lasers operating at 0.633 μm in a ring laser configuration. HeNe lasers are generally present in educational and research optical laboratories.

Applications

Red HeNe lasers have many industrial and scientific uses. They are widely used in laboratory demonstrations in the field of optics because of their relatively low cost and ease of operation compared to other visible lasers producing beams of similar quality in terms of spatial coherence (a single-mode Gaussian beam) and long coherence length (however since about 1990 semiconductor lasers have offered a lower-cost alternative for many such applications). A consumer application of the red HeNe laser is the LaserDisc player, made by Pioneer. The laser is used in the device to read the optical disc.

Ion Laser

An ion laser is a gas laser that uses an ionized gas as its lasing medium. Like other gas lasers, ion lasers feature a sealed cavity containing the laser medium and mirrors forming a Fabry–Pérot resonator. Unlike helium–neon lasers, the energy level transitions that contribute to laser action come from ions. Because of the large amount of energy required to excite the ionic transitions used in ion lasers, the required current is much greater, and as a result all but the smallest ion lasers are water-cooled. A small air-cooled ion laser might produce, for example, 130 milliwatts of output light with a tube current of about 10 amperes and a voltage of 105 volts. Since one ampere times one volt is one watt, this is an electrical power input of about one kilowatt. Subtracting the (desirable) light output of 130 mW from power input, this leaves the large amount of waste heat of nearly one kW. This has to be dissipated by the cooling system. In other words, the power efficiency is very low.

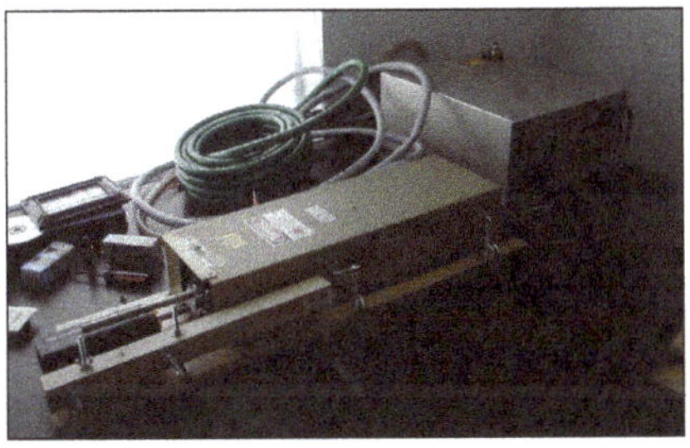

1 mW Uniphase HeNe on alignment rig (left) and 2 W Lexel 88 argon-ion laser (center) with power-supply (right). To the rear are hoses for water cooling.

Types

Krypton Laser

A krypton laser is an ion laser using ions of the noble gas krypton as its gain medium. The laser pumping is done by an electrical discharge. Krypton lasers are widely used in scientific research, and in commercial uses, when the krypton is mixed with argon, it creates a "white-light" lasers, useful for laser light shows. Krypton lasers are also used in medicine (e.g. for coagulation of retina), for the manufacture of security holograms, and numerous other purposes.

Krypton lasers can emit visible light close to several different wavelengths, commonly 406.7 nm, 413.1 nm, 415.4 nm, 468.0 nm, 476.2 nm, 482.5 nm, 520.8 nm, 530.9 nm, 568.2 nm, 647.1 nm, and 676.4 nm.

Argon Laser

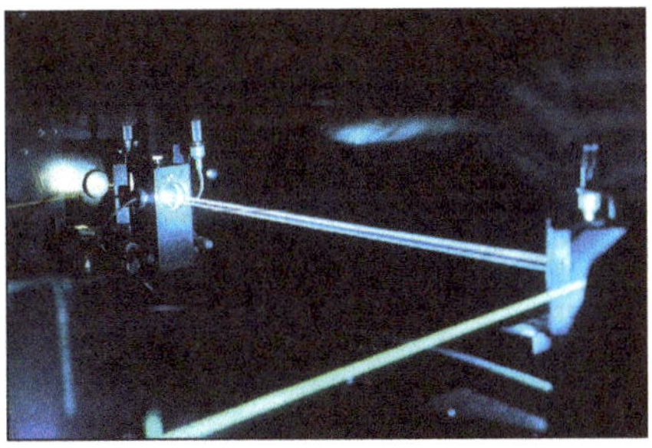

This argon-ion laser emits blue-green light at 488 and 514 nm.

Argon-ion lasers are used for retinal phototherapy (for the treatment of diabetes), lithography, and the pumping of other lasers. Argon-ion lasers emit at 13 wavelengths through the visible and ultraviolet spectra, including: 351.1 nm, 363.8 nm, 454.6 nm, 457.9 nm, 465.8 nm, 476.5 nm, 488.0 nm, 496.5 nm, 501.7 nm, 514.5 nm, 528.7 nm, 1092.3 nm. However, the most commonly used wavelengths are in the blue-green region of the visible spectrum. These wavelengths have the potential for use in underwater communications because seawater is quite transparent in this range of wavelengths.

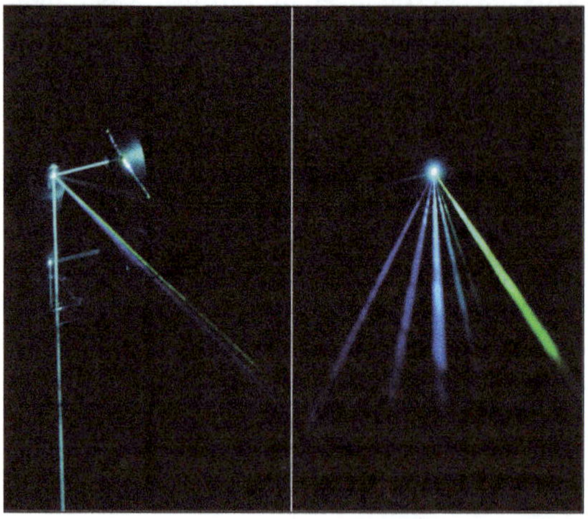

An argon-laser beam consisting of multiple colors (wavelengths) strikes a silicon diffraction mirror grating and is separated into several beams, one for each wavelength (left to right): 458 nm, 476 nm, 488 nm, 497 nm, 502 nm, 515 nm.

Common argon and krypton lasers are capable of emitting continuous-wave (CW) output of several milliwatts to tens of watts. Their tubes are usually made from nickel end bells, kovar metal-to-ceramic seals, beryllium oxide ceramics, or tungsten disks mounted on a copper heat spreader in a ceramic liner. The earliest tubes were simple quartz, then followed by quartz with graphite disks. In comparison with the helium–neon lasers, which require just a few milliamperes of input current, the current used for pumping the krypton laser is several amperes, since the gas has to be ionized. The ion laser tube produces much waste heat, and such lasers require active cooling.

The typical noble-gas ion-laser plasma consists of a high-current-density glow discharge in a noble gas in the presence of a magnetic field. Typical continuous-wave plasma conditions are current densities of 100 to 2000 A/cm^2, tube diameters of 1.0 to 10 mm, filling pressures of 0.1 to 1.0 Torr (0.0019 to 0.019 psi), and an axial magnetic field of the order of 1000 gauss.

William R. Bennett, a co-inventor of the first gas laser (the helium–neon laser), was the first to observe spectral hole burning effects in gas lasers, and he created the theory of "hole burning" effects in laser oscillation. He was co-discoverer of lasers using electron-impact excitation in each of the noble gases, dissociative excitation transfer in the neon–oxygen laser (the first chemical laser), and collision excitation in several metal-vapor lasers.

Commercially Available Types

- Ar/Kr: A mix of argon and krypton can result in a laser with output wavelengths that appear as white light.
- Helium–cadmium: Blue laser emission at 442 nm and ultraviolet at 325 nm.
- Copper vapor: Yellow and green emission at 578 nm and 510 nm.

Experimental

- Xenon,
- Iodine,
- Oxygen.

Application

- Confocal laser scanning microscopy,
- Surgical,
- Laser medicine,
- High speed typesetters,
- Laser light shows,
- DNA sequencers,
- Spectroscopy experiments,
- Pumping dye lasers,
- Semiconductor wafer inspection,
- Direct write high density PCB lithography,
- Fiber Bragg Grating production,
- Long coherence length models can be used for holography.

Nitrogen Laser

A nitrogen laser is a gas laser operating in the ultraviolet range (typically 337.1 nm) using molecular nitrogen as its gain medium, pumped by an electrical discharge.

The wall-plug efficiency of the nitrogen laser is low, typically 0.1% or less, though nitrogen lasers with efficiency of up to 3% have been reported in the literature. The wall-plug efficiency is the product of the following three efficiencies:

- Electrical: TEA laser.

- Gain medium: This is the same for all nitrogen lasers and thus has to be at least 3% .
    - Inversion by electron impact is 10 to 1 due to Franck–Condon principle.
    - Energy lost in the lower laser level: 40%.
    - Optical: More induced emission than spontaneous emission.

A 337nm wavelength and 170 μJ pulse energy 20 Hz cartridge nitrogen laser.

Gain Medium

The gain medium is nitrogen molecules in the gas phase. The nitrogen laser is a three-level laser. In contrast to more typical four-level lasers, the upper laser level of nitrogen is directly pumped, imposing no speed limits on the pump. Pumping is normally provided by direct electron impact; the electrons must have sufficient energy, or they will fail to excite the upper laser level. Typically reported optimum values are in the range of 80 to 100 eV per Torr·cm pressure of nitrogen gas.

There is a 40 ns upper limit of laser lifetime at low pressures and the lifetime becomes shorter as the pressure increases. The lifetime is only 1 to 2 ns at 1 atmosphere.

$$t[\text{ns}] = \frac{36}{1 + 12.8 * p[\text{bar}]}.$$

The strongest lines are at 337.1 nm wavelength in the ultraviolet. Other lines have been reported at 357.6 nm, also ultraviolet. This information refers to the second positive system of molecular nitrogen, which is by far the most common. No vibration of the two nitrogen atoms is involved, because the atom-atom distance does not change with the electronic transition. The rotation needs to change to deliver the angular momentum of the photon, furthermore multiple rotational states are populated at room temperature. There are also lines in the far-red and infrared from the first positive system, and a visible blue laser line from the molecular nitrogen positive (1+) ion.

The metastable lower level lifetime is 40 μs, thus, the laser self-terminates, typically in less than 20 ns. This type of self-termination is sometimes referred to as "bottlenecking in the lower level". This is only a rule of thumb as is seen in many other lasers: The helium–neon laser also has a bottleneck as one decay step needs the walls of the cavity and this laser typically runs in continuous mode. Several organic dyes with upper level lifetimes of less than 10 ns have been used in continuous mode. The Nd:YAG laser has an upper level lifetime of 230 μs, yet it also supports 100 ps pulses.

Repetition rates can range as high as a few kHz, provided adequate gas flow and cooling of the structure are provided. Cold nitrogen is a better medium than hot nitrogen, and this appears to be part of the reason that the pulse energy and power drop as the repetition rate increases to more than a few pulses per second. There are also, apparently, issues involving ions remaining in the laser channel. Air, which is 78% nitrogen, can be used, but more than 0.5% oxygen poisons the laser.

Optics

Nitrogen lasers can operate within a resonator cavity, but due to the typical gain of 2 every 20 mm they more often operate on superluminescence alone; though it is common to put a mirror at one end such that the output is emitted from the opposite end.

For a 10 mm wide gain volume diffraction comes into play after 30 m along the gain medium, a length which is unheard of. Thus this laser does not need a concave lens or refocusing lenses and beam quality improves along the gain medium. The height of the pumped volume may be as small as 1 mm, needing a refocusing lens already after 0.3 m. A simple solution is to use rounded electrodes with a large radius, so that a quadratic pump profile is obtained.

Electrical

The gain medium is usually pumped by a transverse electrical discharge. When the pressure is at (or above) 1013 mbar (atmospheric pressure), the configuration is called a TEA laser Transverse Electrical discharge in gas at Atmospheric pressure, this is also used for pressures down to 30 mbar.

Microscopic Description of a Fast Discharge

In a strong external electric field this electron creates an electron avalanche in the direction of the electric field lines. Diffusion of electrons and elastic scattering at a buffer gas molecule spreads the avalanche perpendicular to the field. Inelastic scattering creates photons, which create new avalanches centimeters away. After some time the electric charge in the avalanche becomes so large that following Coulomb's law it generates an electric field as large as the external electric field. At regions of increased field strength the avalanche effect is enhanced. This leads to electric arc like discharges called streamers. A mix of a noble gas (up to 0.9) and nitrogen enhance elastic scattering of electrons over electron multiplying and thus widens avalanches and streamers.

Spark gaps use a high density of gas molecules and a low density of initial electrons to favor streamers. Electrons are removed by a slowly rising voltage. A high density gas increases the breakdown field, thus shorter arcs can be used with lower inductance and the capacity between the electrodes is increased. A wide streamer has a lower inductance.

Gas lasers use low density of gas molecules and a high density of initial electrons to prevent streamers. Electrons are added by preionisation not removed by oxygen, because nitrogen from bottles is used. Wide avalanches can excite more nitrogen molecules.

Inelastic scattering heats up a molecule, so that in a second scattering the probability of electron emission is increased. This leads to an arc. Typically arcing occurs *after* lasing in nitrogen. The streamer in the spark gap discharges the electrodes only by means of image charge, thus when the streamer touches both electrodes most of the charge is still available to feed the arc; additional charge is stored on the distribution plates. Thus arcing in the spark gap starts *before* lasing.

Electrodynamics

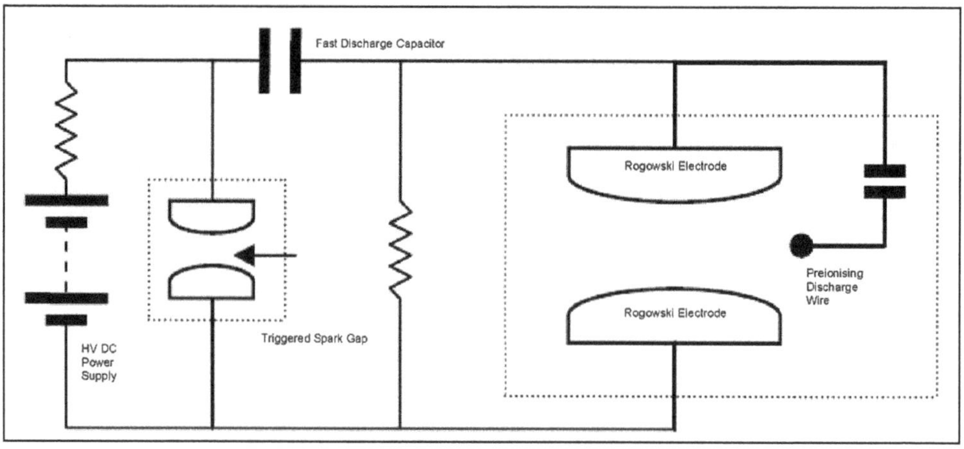

Circuit.

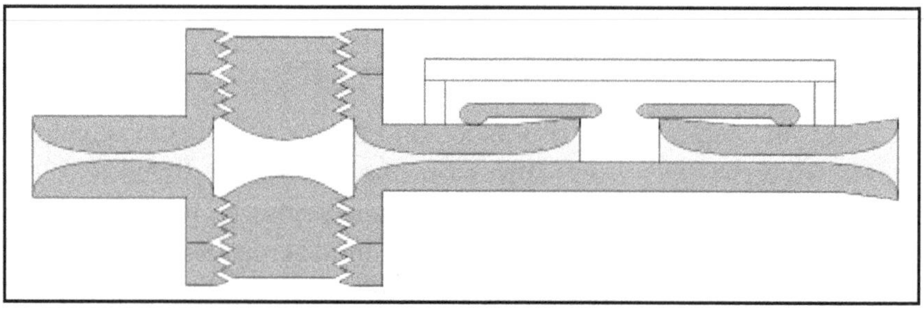

Low inductance implementation cross cut. Erratum:
Right cap needs to be bigger.

The electronics is a circuit composed of a spark gap, a capacitor, and the discharge through the nitrogen. First the spark gap and the capacitor are charged. The spark gap then discharges itself and voltage is applied to the nitrogen.

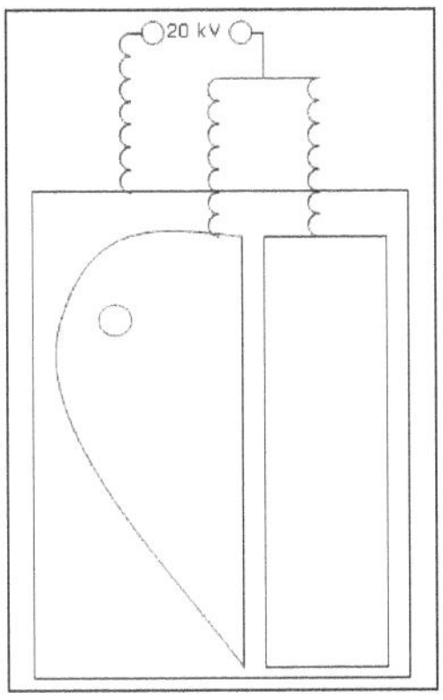

Low inductance implementation top view. Erratum: Caps should be slightly longer than the channel and have rounded corners.

An alternative construction uses two capacitors connected as a Blumlein generator. Two capacitors are connected so that one plate is a common earth, the others are each connected to the spark gap electrodes. These capacitors are often constructed from a single layer of printed circuit board, or similar stack of copper foil and thin dielectric. The capacitors are linked through an inductor, a simple air-spaced coil. One capacitor also has a small spark gap across it. When HT is applied, the two capacitors are charged slowly, effectively linked by the inductor. When the spark gap reaches its triggering voltage, it discharges and quickly reduces that capacitor's voltage to zero. As the discharge is rapid, the inductor acts as an open circuit and so the voltage difference across the transverse spark gap (between the two capacitors) rises rapidly until the main spark gap discharges, firing the laser.

The speed of either circuit is increased in two steps. First, the inductance of all components is reduced by shortening and widening conductors and by squeezing the circuit into a flat rectangle. The total inductance is the sum of the components:

Object	Length	Thickness	Width	Width	Inductance	Inductance	Inductance	Capacity	Oscillation
			As coil	As wire	Measured	Coil theory	Wire theory	Plate theory	Period
Unit	m	m	m	m	nH	nH	nH	nF	ns
Spark gap	2×10^{-2}	1×10^{-2}	2×10^{-2}	1×10^{-5}	10	12.57	13.70	0.0004	
Metal tape	2×10^{-2}	2×10^{-2}	4×10^{-2}	5×10^{-3}		12.57	5.32	0.0004	
Cap. 1	2×10^{-1}	4×10^{-4}	3×10^{-1}			0.34		2.6563	
Metal tape	2×10^{-2}	2×10^{-2}	3×10^{-1}			1.68		0.0027	
Laser channel	1×10^{-2}	2×10^{-2}	3×10^{-1}			0.84		0.0013	
Metal tape	2×10^{-2}	2×10^{-2}	3×10^{-1}			1.68		0.0027	
Cap. 2	3×10^{-1}	4×10^{-4}	3×10^{-1}			0.50		3.9844	
Spark osc.						22.90		2.6563	49
Disch. osc.						5.03		1.5938	18

The intense discharge is reported to distort oscilloscopes nearby. This can be reduced by building the laser symmetrically into a grounded cylinder with the spark gap at the bottom, the laser at the top, capacitor 1 left and right, and capacitor 2 left and right stacked onto capacitor 1. This has the further advantage of reducing the inductance. And this has the disadvantage that the laser channel cannot be inspected for sparks anymore.

Secondly, transmission line theory and waveguide theory is applied to achieve a traveling wave excitation. Measured nitrogen laser pulses are so long that the second step is unimportant. From this analysis it follows that:

- The end mirror and the spark gap are on the same side.
- A long narrow laser at atmospheric pressures is ineffective.

Spark Gap

Paschen's law states that the length of the spark gap is inverse-proportional to the pressure. For a fixed length to diameter ratio of the spark, the inductance is proportional to the length (source , compare with: dipole antenna). Thus the electrodes of the spark gap are glued or welded on a dielectric spacer-ring. To reduce the danger due to the pressure, the volume is minimized. To prevent sparks outside space ring in the low pressure the spacer usually gets thicker outwards in an s-shaped manner.

Connection between spark gap and laser channel based on traveling wave theory:

- The low inductance spark gap may be inserted into a strip transmission line.

The breakdown voltage is low for helium, medium for nitrogen and high for SF_6, though nothing is said about the spark thickness variations.

Rise times as high as 8×10^{10} A/s are possible with a spark gap. This nicely matches the typical rise times of 1×10^{-8} s and typical currents of 1×10^{3} A occurring in nitrogen lasers.

A cascade of spark gaps allows to use a weak trigger pulse to initiate a streamer in the smaller gap, wait for its transition into an arc, and then for this arc to extend into the larger gap. Still the first spark gap in the cascade needs a free electron to start with, so jitter is rather high.

Preionisation

Avalanches homogenize a discharge fast mostly along the field lines. With a short duration (<10 ms) since the last laser pulse enough ions are left over so that all avalanches overlap also laterally. With low pressure (<100 kPa) the max charge carrier density is low and the electromagnetic driven transition from avalanche to spark is inhibited.

In other cases UV radiation homogenizes a discharge slowly perpendicular to a discharge. These are brought into balance by placing two linear discharges next to each other 1 cm apart. The first discharge is across a smaller gap and starts early. Due to the low number of initial electrons streamers typically 1 mm apart are seen. The electrodes for the first discharge are covered by a dielectric, which limits this discharge. Therefore the voltage is able to rise further until avalanches can start in the second gap. These are so many that they overlap and excite every molecule.

With about 11 ns the UV generation, ionisation, and electron capture are in a similar speed regime as the nitrogen laser pulse duration and thus as fast electric must be applied.

Excitation by Electron Impact

The upper laser level is excited efficiently by electrons with more than 11 eV, best energy is 15 eV. The electron temperature in the streamers only reaches 0.7 eV. Helium due to its higher ionisation energy and lack of vibrational excitations increases the temperature to 2.2 eV. Higher voltages increase the temperature. Higher voltages mean shorter pulses.

Typical Devices

The gas pressure in a nitrogen laser ranges from a few mbar to as much as several bar. Air provides significantly less output energy than pure nitrogen or a mixture of nitrogen and helium. The pulse energy ranges from 1 μJ to about 1 mJ; peak powers between 1 kW and 3 MW can be achieved. Pulse durations vary from a few hundred picoseconds (at 1 atmosphere partial pressure of nitrogen) to about 30 nanoseconds at reduced pressure (typically some dozens of Torr), though FWHM pulsewidths of 6 to 8 ns are typical.

Amateur Construction

The transverse discharge nitrogen laser has long been a popular choice for amateur home construction, owing to its simple construction and simple gas handling.

Applications

- Transverse optical pumping of dye lasers.
- Measurement of air pollution (Lidar).
- Matrix-assisted laser desorption/ionization.

Carbon Dioxide Laser

The CO_2 laser (*carbon dioxide laser*) is a molecular gas laser based on a gas mixture as the gain medium, which contains carbon dioxide (CO_2), helium (He), nitrogen (N_2), and possibly some hydrogen (H_2), water vapor and/or xenon (Xe). Such a laser is electrically pumped via a gas discharge, which can be operated with DC current, with AC current (e.g. 20–50 kHz) or in the radio frequency (RF) domain. Nitrogen molecules are excited by the discharge into a metastable vibrational level and transfer their excitation energy to the CO_2 molecules when colliding with them. Helium serves to depopulate the lower laser level and to remove the heat. Other constituents such as hydrogen or water vapor can help (particularly in sealed-tube lasers) to reoxidize carbon monoxide (formed in the discharge) to carbon dioxide.

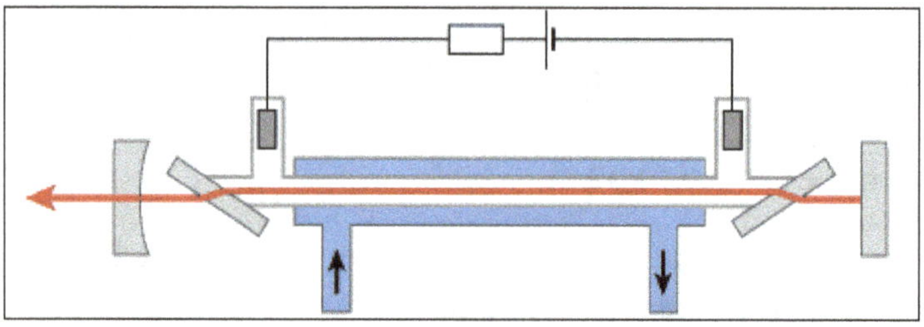

Schematic setup of a sealed-tube carbon dioxide laser. The gas tube has brewster windows and is water-cooled.

CO_2 lasers typically emit at a wavelength of 10.6 μm, but there are other lines in the region of 9–11 μm (particularly at 9.6 μm). In most cases, average powers are between some tens of watts and many kilowatts. The power conversion efficiency can be well above 10%, i.e., it is higher than for most gas lasers (due to a particularly favorable excitation pathway), also higher than for lamp-pumped solid-state lasers, but lower than for many diode-pumped lasers.

Due to their high output powers, CO_2 lasers require high-quality infrared optics, often made of materials like zinc selenide (ZnSe) or zinc sulfide (ZnS).

Laser Types

The family of CO_2 lasers is very diverse:

- For laser powers between a few watts and a several hundred watts, it is common to use sealed-tube or no-flow lasers, where the laser bore and gas supply are contained in a sealed tube. Such lasers are compact and rugged, and reach operation lifetimes of several thousands of hours.

- High-power diffusion-cooled slab lasers have the gas in a gap between a pair of planar water-cooled RF electrodes. The excess heat is efficiently transferred to the electrodes by diffusion, if the electrode spacing is made small compared with the electrode width. Several kilowatts of output are possible.

- Fast axial flow lasers and fast transverse flow lasers are also suitable for multi-kilowatt continuous-wave output powers. The excess heat is removed by the fast-flowing gas mixture, which passes an external cooler before being used again in the discharge.

- Transverse excited atmosphere (TEA) lasers have a very high (about atmospheric) gas pressure. As the voltage required for a longitudinal discharge would be too high, transverse excitation is done with a series of electrodes along the tube. TEA lasers are operated in pulsed mode only, as the gas discharge would not be stable at high pressures. They often produce average output powers below 100 W, but can also be made for powers of tens of kilowatts (combined with high pulse repetition rates).

- There are gas dynamic CO_2 lasers for multi-megawatt powers (e.g. for anti-missile weapons), where the energy is not provided by a gas discharge but by a chemical reaction in a kind of rocket engine.

The concepts differ mainly in the technique of heat extraction, but also in the gas pressure and electrode geometry used. In low-power sealed-tube lasers (used e.g. for laser marking), waste heat is transported to the tube walls by diffusion or a slow gas flow. The beam quality can be very high. High-power CO_2 lasers utilize a fast forced gas convection, which may be in the axial direction (i.e., along the beam direction) or in the transverse direction (for the highest powers).

Applications

CO_2 lasers are widely used for material processing, in particular for:

- Cutting plastic materials, wood, die boards, etc., exhibiting high absorption at 10.6 μm, and requiring moderate power levels of 20–200 W.
- Cutting and welding metals such as stainless steel, aluminum or copper, applying multi-kilowatt powers.
- Laser marking of various materials.

Other applications include laser surgery (including ophthalmology) and range finding. CO_2 lasers used for material processing (e.g. welding and cutting of metals, or laser marking) are in competition with solid-state lasers (particularly YAG lasers and fiber lasers) operating in the 1-μm wavelength regime. These shorter wavelengths have the advantages of more efficient absorption in a metallic workpiece, and the potential for beam delivery via fiber cables. (There are no optical fibers for high-power 10-μm laser beams.) The potentially smaller beam parameter product of 1-μm lasers can also be advantageous. However, the latter potential normally cannot be realized with high-power lamp-pumped lasers, and diode-pumped lasers tend to be more expensive. For these reasons, CO_2 lasers are still widely used in the cutting and welding business, particularly for parts with a thickness greater than a few millimeters, and their sales make more than 10% of all global laser sales (as of 2013). This may to some extent change in the future due to the development of high-power thin-disk lasers and advanced fiber cables in combination with techniques which exploit the high beam quality of such lasers.

Due to their high powers and high drive voltages, CO_2 lasers raise serious issues of laser safety. However, their long operation wavelength makes them relatively eye-safe at low intensities.

Excimer Laser

An excimer laser, sometimes more correctly called an exciplex laser, is a form of ultraviolet laser which is commonly used in the production of microelectronic devices, semiconductor based integrated circuits or "chips", eye surgery, and micromachining.

The excimer laser was invented in 1970 by Nikolai Basov, V. A. Danilychev and Yu. M. Popov, at the Lebedev Physical Institute in Moscow, using a xenon dimer (Xe_2) excited by an electron beam to give stimulated emission at 172 nm wavelength. A later improvement, developed by many groups in 1975 was the use of noble gas halides (originally XeBr). These groups include the Avco Everett Research Laboratory, Sandia Laboratories, the Northrop Research and Technology Center, and the United States Government's Naval Research Laboratory who also developed a XeCl Laser that was excited using a microwave discharge.

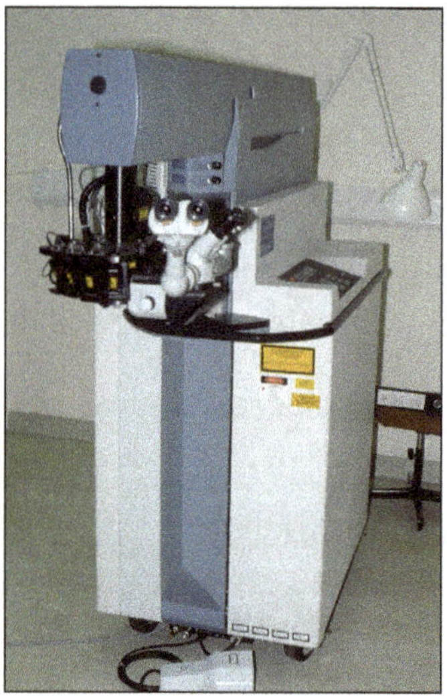

An excimer laser.

Construction

An excimer laser typically uses a combination of a noble gas (argon, krypton, or xenon) and a reactive gas (fluorine or chlorine). Under the appropriate conditions of electrical stimulation and high pressure, a pseudo-molecule called an excimer (or in the case of noble gas halides, exciplex) is created, which can only exist in an energized state and can give rise to laser light in the ultraviolet range.

Operation

Laser action in an excimer molecule occurs because it has a bound (associative) excited state, but a repulsive (dissociative) ground state. Noble gases such as xenon and krypton are highly inert and do not usually form chemical compounds. However, when in an excited state (induced by electrical discharge or high-energy electron beams), they can form temporarily bound molecules with themselves (excimer) or with halogens (exciplex) such as fluorine and chlorine. The excited compound can release its excess energy by undergoing spontaneous or stimulated emission, resulting in a strongly repulsive ground state molecule which very quickly (on the order of a picosecond) dissociates back into two unbound atoms. This forms a population inversion.

Wavelength Determination

The wavelength of an excimer laser depends on the molecules used, and is usually in the ultraviolet:

Excimer	Wavelength	Relative power
Ar_2^*	126 nm	
Kr_2^*	146 nm	
F_2^*	157 nm	
Xe_2^*	172 & 175 nm	
ArF	193 nm	60
KrCl	222 nm	25
KrF	248 nm	100
XeBr	282 nm	
XeCl	308 nm	50
XeF	351 nm	45

Excimer lasers, such as XeF and KrF, can also be made slightly *tunable* using a variety of prism and grating intracavity arrangements.

Repetition Rate

Discharge-pumped excimer lasers are usually operated with a pulse repetition rate of around 100 Hz and a pulse duration of ~10 ns, although some operate at pulse repetition rates as high as 8 kHz and some have pulse length as large as 30 ns. For electron-beam pumped lasers typical pulse length can be as large as 100 ns and a repetition rate usually limited to a single shot within several minutes, although some operate at pulse repetition rates as high as 10 Hz.

Major Applications

Photolithography

Excimer lasers are widely used in high-resolution photolithography machines, one of the critical technologies required for microelectronic chip manufacturing. Current state-of-the-art lithography tools use deep ultraviolet (DUV) light from the KrF and ArF excimer lasers with wavelengths of 248 and 193 nanometers (the dominant lithography technology today is thus also called "excimer laser lithography"), which has enabled transistor feature sizes to shrink below 45 nanometers. Excimer laser lithography has thus played a critical role in the continued advance of the so-called Moore's law for the last 25 years.

The most widespread industrial application of excimer lasers has been in deep-ultraviolet photolithography, a critical technology used in the manufacturing of microelectronic devices (i.e., semiconductor integrated circuits or "chips"). Historically, from the early 1960s through the mid-1980s, mercury-xenon lamps had been used in lithography for their spectral lines at 436, 405 and 365 nm wavelengths. However, with the semiconductor industry's need for both higher resolution (to produce denser and faster chips) and higher throughput (for lower costs), the lamp-based lithography tools were no longer able to meet the industry's requirements. This challenge was overcome when in a pioneering development in 1982, deep-UV excimer laser lithography was proposed and demonstrated at IBM by Kanti Jain. With phenomenal advances made in equipment technology in the last two decades, and today microelectronic devices fabricated using excimer laser lithography totaling $400 billion in annual production, it is the semiconductor industry view that excimer laser lithography has been a crucial factor in the continued advance of Moore's law, enabling minimum features sizes in chip manufacturing to shrink from 800 nanometers in 1990 to 10 nanometers in 2016. From an even broader scientific and technological perspective, since the invention of the laser in 1960, the development of excimer laser lithography has been highlighted as one of the major milestones in the 50-year history of the laser.

Medical Uses

The ultraviolet light from an excimer laser is well absorbed by biological matter and organic compounds. Rather than burning or cutting material, the excimer laser adds enough energy to disrupt the molecular bonds of the surface tissue, which effectively disintegrates into the air in a tightly controlled manner through ablation rather than burning. Thus excimer lasers have the useful property that they can remove exceptionally fine layers of surface material with almost no heating or change to the remainder of the material which is left intact. These properties make excimer lasers well suited to precision micromachining organic material (including certain polymers and plastics), or delicate surgeries such as eye surgery LASIK. Intrigued, they investigated further, finding that the laser made clean, precise cuts that would be ideal for delicate surgeries. This resulted in a fundamental patent and Srinivasan, Blum and Wynne were elected to the National Inventors Hall of Fame in 2002. In 2012, the team members were honored with National Medal of Technology and Innovation by the President of The United States Barack Obama for their work related to the excimer laser. Subsequent work introduced the excimer laser for use in angioplasty. Xenon chloride (308 nm) excimer lasers can also treat a variety of dermatological conditions including psoriasis, vitiligo, atopic dermatitis, alopecia areata and leukoderma.

As light sources, excimer lasers are generally large in size, which is a disadvantage in their medical applications, although their sizes are rapidly decreasing with ongoing development.

Research is being conducted to compare differences in safety and effectiveness outcomes between conventional excimer laser refractive surgery and wavefront-guided or wavefront-optimized refractive surgery, as wavefront methods may better correct for higher-order aberrations.

Scientific Research

Excimer lasers are also widely used in numerous fields of scientific research, both as primary sources and, particularly the XeCl laser, as pump sources for tunable dye lasers, mainly to excite laser dyes emitting in the blue-green region of the spectrum. These lasers are also commonly used in Pulsed laser deposition systems, where their large fluence, short wavelength and non-continuous beam properties make them ideal for the ablation of a wide range of materials.

References

- Hoffman Toschek, et al., "The Pulsed Xenon Ion Laser: Covers the UV, visible, and near-IR with optics changes", IEEE Journal of Quantum Electronics
- Gas-lasers: rp-photonics.com, Retrieved 13 April, 2019
- Csele, Mark (2004). "The TEA Nitrogen Gas Laser". Homebuilt Lasers Page. Archived from the original on 2007-09-11. Retrieved 2007-09-15
- Lasers: rp-photonics.com, Retrieved 15 June, 2019
- Wolford, M. F.; Hegeler, F.; Myers, M. C.; Giuliani, J. L.; Sethian, J. D. (2004). "Electra: Repetitively pulsed, 500 J, 100 ns, KRF oscillator". Applied Physics Letters. 84 (3): 326–328. Bibcode:2004ApPhL..84..326W. doi:10.1063/1.1641513

Solid State Lasers 4

- **Ruby Laser**
- **Nd:YAG Laser**
- **Er:YAG Laser**
- **Diode-pumped Solid-state Laser**

Solid-state laser is a type of laser that makes use of gain medium which is a solid. A few types of solid state lasers are ruby laser, Nd: YAG laser, Er:YAG laser and diode-pumped solid-state laser. The topics elaborated in this chapter will help in gaining a better perspective about these types of solid-state lasers.

The lasers having the solid material as an active medium is called the solid state lasers. Generally there are two classes of solid state lasers.

Continuous Wave Types

Pulsed Solid State Lasers

The output characteristics of the two are different while the construction and function of all solid state lasers is same.

Characteristics of Solid State Lasers

Continuous Wave Type

The term Continuous wave or CW normally indicates that a laser has continuous output. Usually in CW case of laser is continuously pumped.

There are very few solid state crystals that can produce laser light and resist the extreme heat generated by a CW pumping source. Only common laser of this type is the Nd:YAG. Output wave of the Nd:YAG laser is 1.06 microns which is in the near infrared spectrum. The beam diameter varies from 0.75mm to 6mm while beam profile can e either TEM0,0 or multimode. The TEM0,0 beams tend to have narrower beam diameters.

Importantly not that the beam divergence from a solid state laser is not constant. As the laser rod gets heated by the light source during pumping, thermal expansion occurs.

This causes the rod to act as a lens and expand beam. So during use when the rod gets hotter than the rod expands even more and so causing greater divergence.

The beam divergence for the CW solid state laser can be as low as 1 milli radian for TEM0,0 mode lasers are as high as 20 milli radians fro multimode lasers. The output power for the CW solid state laser varies from a low of 0.4 watt to a high of 600watts.

Since a Q-switch can be inserted between the laser mirrors, therefore a very rapidly pulsed output can be obtained. In addition the CW pumped solid state lasers can be mode locked for the production of ultra short pulses. In fact these lasers can produce a CW, modulated CW, Q-switched or mode locked output.

Pulsed Type

The pulsed solid state laser produces a pulsed output due to the pulsing of the input energy. In pulsed lasers the material used can be cooled between pulses therefore the active medium does not exposed to the extreme temperature rise, experienced by a CW laser. That is why there are more choices of active media for pulsed laser rather than for CW solid state lasers.

When Nd:YAG is used as a active medium then it can pulsed very rapidly because it can even with stand the extreme thermal loads of CW operation. It is very common to pulse YAG lasers a rate of up to 200 pulses per second. However if ruby or Nd: glass is used as the active medium then pulse rates are limited to 2 or 3 pulses per second. The additional time between pulses is needed to cool the crystal to prevent fracturing.

The Nd:YAG and ND: glass put out the wavelengths in the near infrared spectrum, i.e. 1604 and 1600nm respectively. Ruby laser emits a visible beam at 694nm (lightly red).

Beam diameter of pulsed laser is in 5 to 10mm range. Beam divergence ranges from 1 milli radian to 10 milli radian. Power output from pulsed laser averages about 400watts although the peak power of individual pulse is much higher.

Construction of Solid State Lasers

The typical solid state laser construction can be shown as:

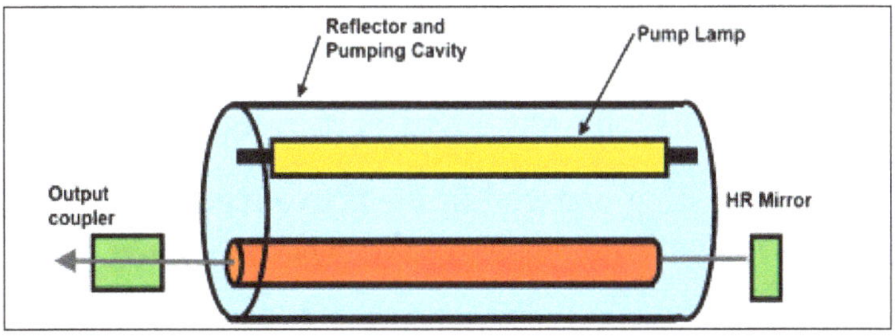

Usually all solid state lasers have similar design. A laser rod is mounted near an arc or flash lamp. The lamp is connected to DC power supply that maintains a controlled current through lamp. The laser rod and lamp are placed parallel to each other and are surrounded by a reflector. The high reflective (HR) mirror and the output coupler are placed at either ends of the laser cavity. For simplicity we do not show the DC power supply and cooling system, however they exists in solid state lasers.

Function of Solid State Lasers

We know that active medium used for solid state lasers is a solid material. Usually all solid state lasers are pumped optically means light source is used as energy source of solid state lasers. When the solid material rod absorbs the light energy from light source then becomes excited. The upper energy level of the ions are radiation less, however when the energy transition takes place then the meta stable upper laser level is soon reached. At this point the emissions occurs which obviously results in lasing.

By energy level diagram the function of solid state say Nd:YAG laser can be shown as:

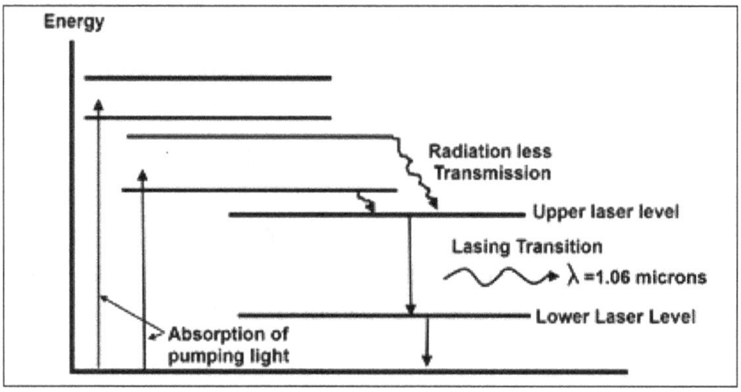

Applications/Uses of Solid State Lasers

- Nd:YAG solid state lasers usually used when drilling holes in metals.
- Nd:YAG pulsed type solid state lasers can be used in medical applications such as in endoscopy etc.
- As military application, Nd:YAG is used by target destination system.

Advantages of Solid State Lasers

The solid state lasers have the following advantages over other types of lasers.

- No chance is of wasting material in the active medium because here material used is in solid form not in gas form, where this occurs.
- Both continuous and pulsed output is possible from solid state lasers.
- Solid state lasers have high efficiency from some of gas lasers such as He-Ne lasers and Argon Lasers efficiency of solid state Nd:YAG laser is 2% to 3%.

- Construction of solid state laser is comparatively simple. Beam diameter of solid state laser is very less than CO_2 lasers.
- Output power ranging from very low value of about 0.04 watts to high value of about 600 watts.
- Cast of solid state lasers is economical.

Disadvantages of Solid State Lasers

- Efficiency of solid state laser is very low as compared to CO_2 lasers.
- Great disadvantage of solid state lasers is the divergence, which is not constant and ranges 1 milli radian to 20 milli radian.
- Output power is also not very high as in CO_2 lasers.
- Due to thermal lasing in solid state lasers, the power loss occurs when the rod gets too hot.

Producing Solid State Rods for Lasers

The active medium in solid state lasers can be one of different crystals. These crystals are not found in nature but rather they are produced commercially such as Ruby, Nd:YAG (Neodymium: Yttrium, Aluminum garnet), Nd: glass (Neodymium: glass), erbium etc.

The crystals used in the lasers are made by doping a highly transparent host material with a metal that will lase. For example YAG host material that has been heated and is in melton form can be doped with Nd. When the mixture cools, a crystal begins to form. A cylindrical crystal is then carefully drawn from the melton material when it continues to cool. This process is called growing a crystal.

Once the crystal is completely grown then ends of the cylinder are polished to perfection. The final result is a Nd:YAG rod, that is used as an active medium in a solid state laser. Other solid state lasers rods are produced in similar manner.

Above process that any variation in temperature of the melton material can cause distortion in the crystal. The Nd comprises between 0.7 & 1.25 of the total weight of the Nd:YAG active medium. It must be evenly distributed throughout the crystal or "hot spots" will appear in crystal. These two can cause distortion.

Ruby Laser

A ruby laser is a solid-state laser that uses the synthetic ruby crystal as its laser medium. Ruby laser is the first successful laser developed by Maiman in 1960.

Ruby laser is one of the few solid-state lasers that produce visible light. It emits deep red light of wavelength 694.3 nm.

Construction of Ruby Laser

A ruby laser consists of three important elements: laser medium, the pump source, and the optical resonator.

Laser Medium or Gain Medium in Ruby Laser

In a ruby laser, a single crystal of ruby (Al_2O_3: Cr_3^+) in the form of cylinder acts as a laser medium or active medium. The laser medium (ruby) in the ruby laser is made of the host of sapphire (Al_2O_3) which is doped with small amounts of chromium ions (Cr_3^+). The ruby has good thermal properties.

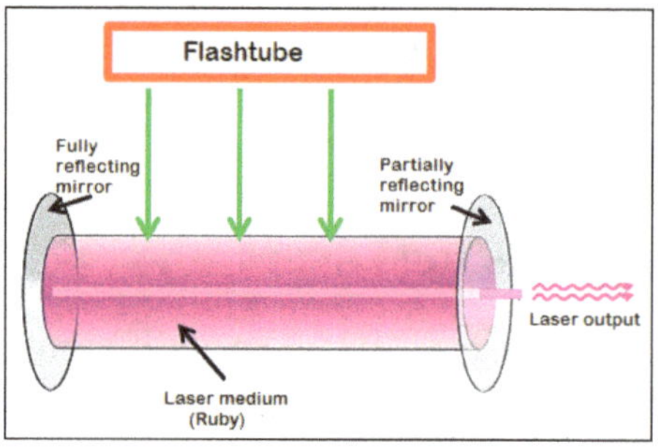

Pump Source or Energy Source in Ruby Laser

The pump source is the element of a ruby laser system that provides energy to the laser medium. In a ruby laser, population inversion is required to achieve laser emission. Population inversion is the process of achieving the greater population of higher energy state than the lower energy state. In order to achieve population inversion, we need to supply energy to the laser medium (ruby).

In a ruby laser, we use flashtube as the energy source or pump source. The flashtube supplies energy to the laser medium (ruby). When lower energy state electrons in the laser medium gain sufficient energy from the flashtube, they jump into the higher energy state or excited state.

Optical Resonator

The ends of the cylindrical ruby rod are flat and parallel. The cylindrical ruby rod is placed between two mirrors. The optical coating is applied to both the mirrors. The process of depositing thin layers of metals on glass substrates to make mirror surfaces is called silvering. Each mirror is coated or silvered differently. At one end of the rod, the mirror is fully silvered whereas, at another end, the mirror is partially silvered.

The fully silvered mirror will completely reflect the light whereas the partially silvered mirror will reflect most part of the light but allows a small portion of light through it to produce output laser light.

Working of Ruby Laser

The ruby laser is a three level solid-state laser. In a ruby laser, optical pumping technique is used to supply energy to the laser medium. Optical pumping is a technique in which light is used as energy source to raise electrons from lower energy level to the higher energy level.

Consider a ruby laser medium consisting of three energy levels E_1, E_2, E_3 with N number of electrons. We assume that the energy levels will be $E_1 < E_2 < E_3$. The energy level E_1 is known as ground state or lower energy state, the energy level E_2 is known as metastable state, and the energy level E_3 is known as pump state.

Let us assume that initially most of the electrons are in the lower energy state (E_1) and only a tiny number of electrons are in the excited states (E_2 and E_3). When light energy is supplied to the laser medium (ruby), the electrons in the lower energy state or ground state (E_1) gains enough energy and jumps into the pump state (E_3).

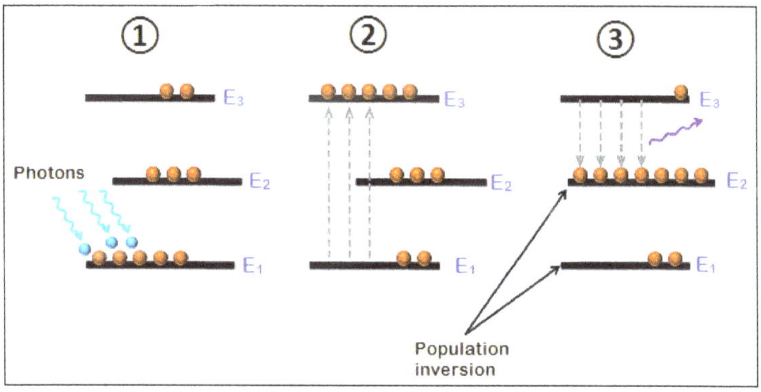

The lifetime of pump state E_3 is very small (10^{-8} sec) so the electrons in the pump state do not stay for long period. After a short period, they fall into the metastable state E_2 by releasing radiationless energy. The lifetime of metastable state E_2 is 10^{-3} sec which is much greater than the lifetime of pump state E_3. Therefore, the electrons reach E_2 much faster than they leave E_2. This results in an increase in the number of electrons in the metastable state E_2 and hence population inversion is achieved.

After some period, the electrons in the metastable state E_2 falls into the lower energy state E_1 by releasing energy in the form of photons. This is called spontaneous emission of radiation.

When the emitted photon interacts with the electron in the metastable state, it forcefully makes that electron fall into the ground state E_1. As a result, two photons are emitted. This is called stimulated emission of radiation.

When these emitted photons again interacted with the metastable state electrons, then 4 photons are produced. Because of this continuous interaction with the electrons, millions of photons are produced.

In an active medium (ruby), a process called spontaneous emission produces light. The light produced within the laser medium will bounce back and forth between the two mirrors. This stimulates other electrons to fall into the ground state by releasing light energy. This is called stimulated emission. Likewise, millions of electrons are stimulated to emit light. Thus, the light gain is achieved. The amplified light escapes through the partially reflecting mirror to produce laser light.

Nd:YAG Laser

Neodymium-doped Yttrium Aluminum Garnet (Nd:YAG) laser is a solid state laser in which Nd:YAG is used as a laser medium.

These lasers have many different applications in the medical and scientific field for processes such as Lasik surgery and laser spectroscopy.

Nd:YAG laser is a four-level laser system, which means that the four energy levels are involved in laser action. These lasers operate in both pulsed and continuous mode.

Nd:YAG laser generates laser light commonly in the near-infrared region of the spectrum at 1064 nanometers (nm). It also emits laser light at several different wavelengths including 1440 nm, 1320 nm, 1120 nm, and 940 nm.

Nd:YAG Laser Construction

Nd:YAG laser consists of three important elements: an energy source, active medium, and optical resonator.

Energy Source

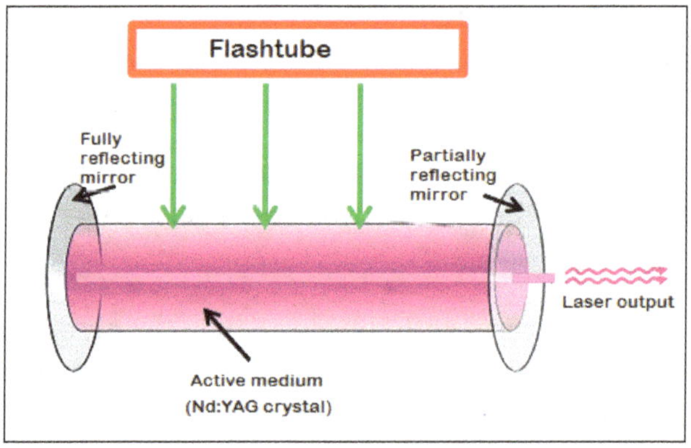

The energy source or pump source supplies energy to the active medium to achieve population inversion. In Nd:YAG laser, light energy sources such as flashtube or laser diodes are used as energy source to supply energy to the active medium.

In the past, flashtubes are mostly used as pump source because of its low cost. However, nowadays, laser diodes are preferred over flashtubes because of its high efficiency and low cost.

Active Medium

The active medium or laser medium of the Nd:YAG laser is made up of a synthetic crystalline material (Yttrium Aluminum Garnet (YAG)) doped with a chemical element (neodymium (Nd)). The lower energy state electrons of the neodymium ions are excited to the higher energy state to provide lasing action in the active medium.

Optical Resonator

The Nd:YAG crystal is placed between two mirrors. These two mirrors are optically coated or silvered.

Each mirror is silvered or coated differently. One mirror is fully silvered whereas, another mirror is partially silvered. The mirror, which is fully silvered, will completely reflect the light and is known as fully reflecting mirror.

On the other hand, the mirror which is partially silvered will reflect most part of the light but allows a small portion of light through it to produce the laser beam. This mirror is known as a partially reflecting mirror.

Working of Nd:YAG Laser

Nd:YAG laser is a four-level laser system, which means that the four energy levels are involved in laser action. The light energy sources such as flashtubes or laser diodes are used to supply energy to the active medium.

In Nd:YAG laser, the lower energy state electrons in the neodymium ions are excited to the higher energy state to achieve population inversion.

Consider a Nd:YAG crystal active medium consisting of four energy levels E_1, E_2, E_3, and E_4 with N number of electrons. The number of electrons in the energy states E_1, E_2, E_3, and E_4 will be N1, N_2, N_3, and N_4.

Let us assume that the energy levels will be $E_1 < E_2 < E_3 < E_4$. The energy level E_1 is known as ground state, E_2 is the next higher energy state or excited state, E3 is the metastable state or excited state and E_4 is the pump state or excited state. Let us assume that initially, the population will be $N_1 > N_2 > N_3 > N_4$.

When flashtube or laser diode supplies light energy to the active medium (Nd:YAG crystal), the lower energy state (E_1) electrons in the neodymium ions gains enough energy and moves to the pump state or higher energy state E_4.

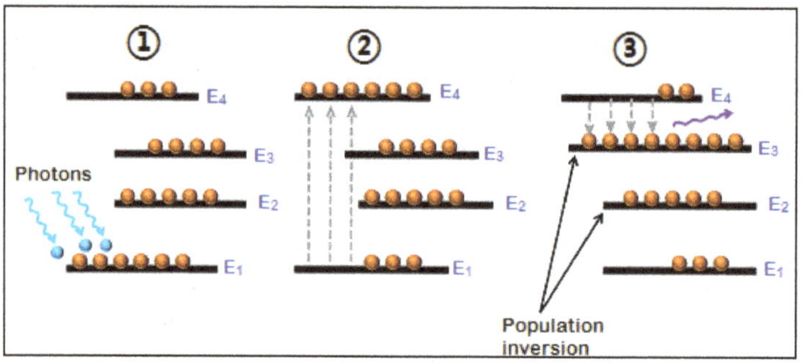

The lifetime of pump state or higher energy state E_4 is very small (230 microseconds (Âμs)) so the electrons in the energy state E_4 do not stay for long period. After a short period, the electrons will fall into the next lower energy state or metastable state E_3 by releasing non-radiation energy (releasing energy without emitting photons).

The lifetime of metastable state E_3 is high as compared to the lifetime of pump state E_4. Therefore, the electrons reach E_3much faster than they leave E_3. This results in an increase in the number of electrons in the metastable E_3 and hence population inversion is achieved.

After some period, the electrons in the metastable state E_3 will fall into the next lower energy state E_2 by releasing photons or light. The emission of photons in this manner is called spontaneous emission.

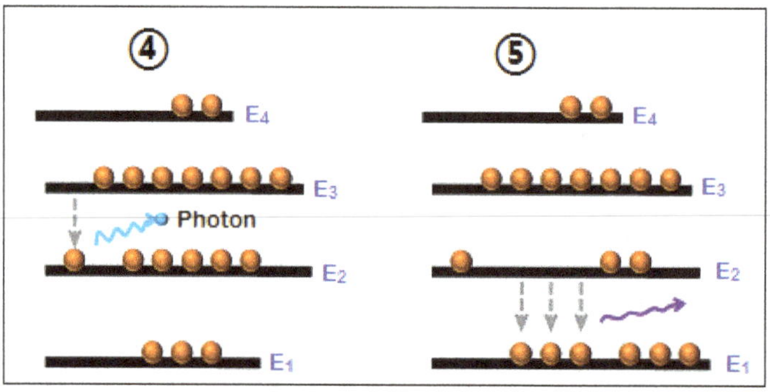

The lifetime of energy state E_2 is very small just like the energy state E_4. Therefore, after a short period, the electrons in the energy state E_2 will fall back to the ground state E_1 by releasing radiationless energy.

When photon emitted due to spontaneous emission is interacted with the other metastable state electron, it stimulates that electron and makes it fall into the lower energy state by releasing the photon. As a result, two photons are released. The emission of photons in this manner is called stimulated emission of radiation.

When these two photons again interacted with the metastable state electrons, four photons are released. Likewise, millions of photons are emitted. Thus, optical gain is achieved.

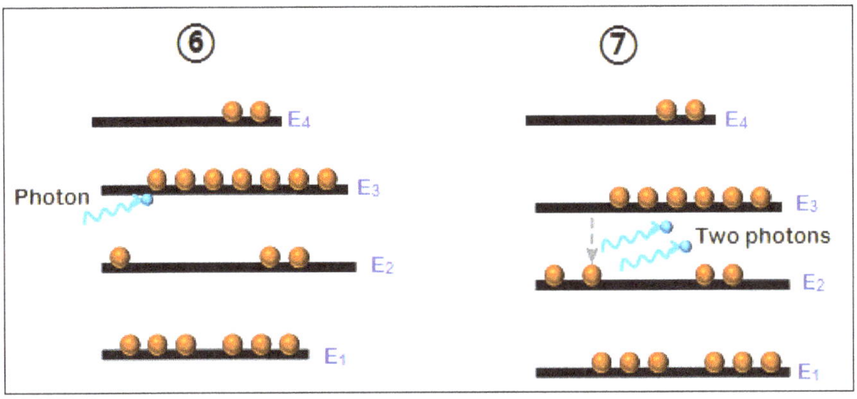

Spontaneous emission is a natural process but stimulated emission is not a natural process. To achieve stimulated emission, we need to supply external photons or light to the active medium.

The Nd:YAG active medium generates photons or light due to spontaneous emission. The light or photons generated in the active medium will bounce back and forth between the two mirrors. This stimulates other electrons to fall into the lower energy state by releasing photons or light. Likewise, millions of electrons are stimulated to emit photons.

The light generated within the active medium is reflected many times between the mirrors before it escapes through the partially reflecting mirror.

Advantages of Nd:YAG laser:

- Low power consumption.
- Nd:YAG laser offers high gain.
- Nd:YAG laser has good thermal properties.
- Nd:YAG laser has good mechanical properties.
- The efficiency of Nd:YAG laser is very high as compared to the ruby laser.

Applications of Nd:YAG Laser

- Military: Nd:YAG lasers are used in laser designators and laser rangefinders. A laser designator is a laser light source, which is used to target objects for attacking. A laser rangefinder is a rangefinder, which uses a laser light to determine the distance to an object.

- Medicine: Nd:YAG lasers are used to correct posterior capsular opacification (a condition that may occur after a cataract surgery). Nd:YAG lasers are used to remove skin cancers.

- Manufacturing: Nd:YAG lasers are used for etching or marking a variety of plastics and metals. Nd:YAG lasers are used for cutting and welding steel.

Er:YAG Laser

Er:YAG lasers are solid-state lasers that have erbium-doped yttrium aluminium garnet as a lasing medium. These lasers emit infrared light at a wavelength of 2940 nm. The energy produced by the Er:YAG laser is 10-15 times more absorbed by water in the skin than the energy from CO_2 lasers. This property limits the applications of Er:YAG laser in surgery and other application where water is present. Thus, this laser is less common than the other related lasers such as Nd:YAG and Er:glass lasers.

Low doped Er:YAG laser is an efficient tool for generating high power and high energy laser emission through two-level resonant pumping mechanisms. Er:YAG laser is commonly used in medical and dental applications.

Laser Properties

Laser Properties	
Laser type	Solid
Pump source	Laser diode
Operating wavelength	1.53-1.56 μm

Physical and Chemical Properties

Physical and Chemical Properties	
Chemical formula	$Er^{3+}:Y_3Al_5O_{12}$
Molecular weight	593.7 g mol^{-1}
Crystal structure	Cubic
Melting point	1965 °C
Mohs hardness	8.25
Thermal conductivity	11.2 W m^{-1} K^{-1}
Specific heat capacity	0.59 J g^{-1} K^{-1}
Young's modulus	335 GPa

Applications

Er:YAG lasers have been widely used for laser resurfacing of human skin that includes melasma, deep rhytides and acne scarring. The energy of Er:YAG lasers is also absorbed well by hydroxyapatite in addition to water, and hence they are suitable for cutting bone as well as soft tissues.

They are effective for treating tooth decay atraumatically, without the use of local anesthetia to numb the tooth. They eliminate the vibration of the dental drill to avoid the risk of causing microfractures in the tooth. However, the laser energy provides a sedative effect on the nerve when used at low power settings thereby creating a sensation of pain in the tooth.

Other major applications of Er:YAG laser include:

- Dentistry,
- Removal of warts,
- Drug delivery.

Diode-pumped Solid-state Laser

Diode-pumped solid-state lasers (DPSSLs) are solid-state lasers made by pumping a solid gain medium, for example, a ruby or a neodymium-doped YAG crystal, with a laser diode.

DPSSLs have advantages in compactness and efficiency over other types, and high power DPSSLs have replaced ion lasers and flashlamp-pumped lasers in many scientific applications, and are now appearing commonly in green and other color laser pointers.

Coupling

The wavelength of laser diodes is tuned by means of temperature to produce an optimal compromise between the absorption coefficient in the crystal and energy efficiency (lowest possible pump photon energy). As waste energy is limited by the thermal lens this means higher power densities compared to high-intensity discharge lamps.

High power lasers use a single crystal, but many laser diodes are arranged in strips (multiple diodes next to each other in one substrate) or stacks (stacks of substrates). This diode grid can be imaged onto the crystal by means of a lens. Higher brightness (leading to better beam profile and longer diode lifetimes) is achieved by optically removing the dark areas between the diodes, which are needed for cooling and delivering the current. This is done in two steps:

- The "fast axis" is collimated with an aligned grating of cylindrical micro-lenses.
- The partially collimated beams are then imaged at reduced size into the crystal. The crystal can be pumped longitudinally from both end faces or transversely from three or more sides.

The beams from multiple diodes can also be combined by coupling each diode into an optical fiber, which is placed precisely over the diode (but behind the micro-lens). At the other end of the fiber bundle, the fibers are fused together to form a uniform, gap-less, round profile on the crystal. This also permits the use of a remote power supply.

Some Numbers

High power laser diodes are fabricated as bars with multiple single strip laser diodes next to each other.

Each single strip diode typically has an active volume of:

1 μm	2 mm	100 μm
Height	Depth	Width
Fast axis	Optical axis	Slow axis

and depending on the cooling technique for the whole bar (100 to 200) μm distance to the next laser diode.

The end face of the diode along the fast axis can be imaged onto strip of 1 μm height. But the end face along the slow axis can be imaged onto a smaller area than 100 μm. This is due to the small divergence (hence the name: 'slow axis') which is given by the ratio of depth to width. Using the above numbers the fast axis could be imaged onto a 5 μm wide spot.

So to get a beam which is equal divergence in both axis, the end faces of a bar composed of 5 laser diodes, can be imaged by means of 4 (acylindrical) cylinder lenses onto an image plane with 5 spots each with a size of 5 mm x 1 mm. This large size is needed for low divergence beams. Low divergence allows paraxial optics, which is cheaper, and which is used to not only generate a spot, but a long beam waist inside the laser crystal (length = 50 mm), which is to be pumped through its end faces.

Also in the paraxial case it is much easier to use gold or copper mirrors or glass prisms to stack the spots on top of each other, and get a 5 x 5 mm beam profile. A second pair of (spherical) lenses image this square beam profile inside the laser crystal.

In conclusion a volume of 0.001 mm³ active volume in the laser diode is able to saturate 1250 mm³ in a $Nd:YVO_4$ crystal.

Common DPSSL Processes

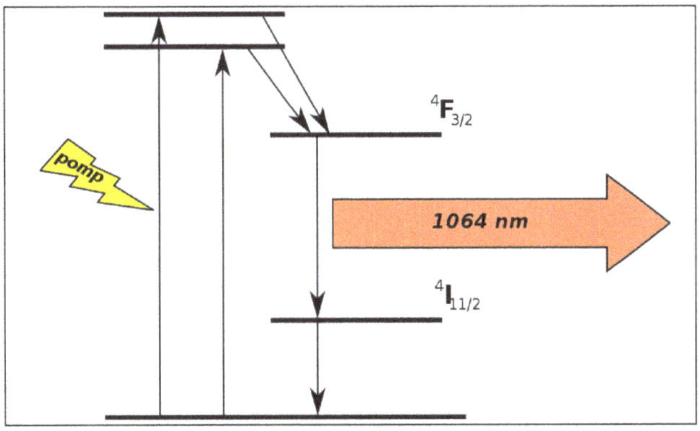

Neodymium ions in various types of ionic crystals, and also in glasses, act as a laser gain medium, typically emitting 1,064 nm light from a particular atomic transition in the neodymium ion, after being "pumped" into excitation from an external source.

The most common DPSSL in use is the 532 nm wavelength green laser pointer. A powerful (>200 mW) 808 nm wavelength infrared GaAlAs laser diode pumps a neodymium-doped yttrium aluminium garnet (Nd:YAG) or a neodymium-doped yttrium orthovanadate ($Nd:YVO_4$) crystal which produces 1064 nm wavelength light from the main spectral transition of neodymium ion. This light is then *frequency doubled* using a nonlinear optical process in a KTP crystal, producing 532 nm light. Green DPSSLs are usually around 20% efficient, although some lasers can reach up to 35% efficiency. In other words, a green DPSSL using a 2.5 W pump diode would be expected to output around 500-900 mW of 532 nm light.

In optimal conditions, $Nd:YVO_4$ has a conversion efficiency of 60%, while KTP has a conversion efficiency of 80%. In other words, a green DPSSL can theoretically have an overall efficiency of 48%.

In the realm of very high output powers, the KTP crystal becomes susceptible to optical damage. Thus, high-power DPSSLs generally have a larger beam diameter, as the 1064 nm laser is expanded before it reaches the KTP crystal, reducing the irradiance from the infrared light. In order to maintain a lower beam diameter, a crystal with a higher damage threshold, such as LBO, is used instead.

Blue DPSSLs use a nearly identical process, except that the 808 nm light is being converted by an Nd:YAG crystal to 946 nm light (selecting this non-principal spectral line of neodymium in the same Nd-doped crystals), which is then frequency-doubled to 473 nm by a beta barium borate (BBO) or lithium triborate (LBO) crystal. Because of the lower gain for the materials, blue lasers are relatively weak, and are only around 3-5% efficient. In the late 2000s, it was discovered that bismuth triborate (BiBO) crystals were more efficient than BBO and LBO and do not have the disadvantage of being hygroscopic, which degrades the crystal if it is exposed to moisture.

Yellow DPSSLs use an even more complicated process: An 808 nm pump diode is used to generate 1,064 nm and 1,342 nm light, which are summed in parallel to become 593.5 nm. Due to their complexity, most yellow DPSSLs are only around 1% efficient, and usually more expensive per unit of power.

Another method is to generate 1,064 and 1,319 nm light, which are summed to 589 nm. This process is more efficient, with about 3% of the pump diode's power being converted to yellow light.

Comparison to Diode Lasers

DPSSLs and diode lasers are two of the most common types of solid-state lasers. However, both types have their advantages and disadvantages.

DPSSLs generally have a higher beam quality and can reach very high powers while maintaining a relatively good beam quality. Because the crystal pumped by the diode acts as its own laser, the quality of the output beam is independent of that of the input beam. In comparison, diode lasers can only reach a few hundred milliwatts unless they operate in multiple transverse mode. Such multi-mode lasers have a larger beam diameter and a greater divergence, which often makes them less desirable. In fact, single-mode operation is essential in some applications, such as optical drives.

On the other hand, diode lasers are cheaper and more energy efficient. As DPSSL crystals are not 100% efficient, some power is lost when the frequency is converted. DPSSLs are also more sensitive to temperature and can only operate optimally within a small range. Otherwise, the laser would suffer from stability issues, such as hopping between modes and large fluctuations in the output power. DPSSLs also require a more complex construction.

Diode lasers can also be precisely modulated with a greater frequency than DPSSLs. Neodymium-doped solid state lasers continue to be the laser source of choice for industrial applications. Direct pumping of the upper Nd laser level at 885-nm (rather than at the more traditional broad 808-nm band) offers the potential of improved performance through a reduction in the lasing quantum defect, thereby improving system efficiency, reducing cooling requirements, and enabling further TEM00 power scaling. Because of the narrow 885-nm absorption feature in Nd:YAG, certain systems may benefit from the use of wavelength-locked diode pump sources, which serve to narrow and stabilize the pump emission spectrum to keep it closely aligned to this absorption feature. To date, high power diode laser locking schemes such as internal distributed feedback Bragg gratings and externally aligned volume holographic grating optics, VHG's, have not been widely implemented due to the increased cost and assumed performance penalty of the technology. However, recent advancements in the manufacture of stabilized diode pump sources which utilize external wavelength locking now offer improved spectral properties with little-to-no impact on power and efficiency. Benefits of this approach include improvements in laser efficiency, spectral linewidth, and pumping efficiency.

References

- Solid-state-lasers, microwave-radar, electronics: daenotes.com, Retrieved 16 July, 2019
- Rubylaserdefinitionconstructionworking, laser, physics, physics-and-radio-electronics.com, Retrieved 17 August, 2019
- Leisher, Paul. "Commercial High-Efficiency 885-nm Diode Lasers" (PDF). Nlight. Retrieved 18 May 2012
- Ndyaglaser, laser, physics: physics-and-radio-electronics.com, Retrieved 19 January, 2019

Semiconductor Lasers 5

- Laser Diode
- Distributed Feedback Laser
- Vertical-cavity Surface-emitting Laser
- Quantum Cascade Laser
- Interband Cascade Laser
- Quantum Well Laser
- Vertical-external-cavity Surface-emitting-laser
- Distributed Bragg Reflector Laser
- Quantum Dot Laser
- Transistor Laser

Semiconductor lasers consist of a semiconductor diode which produces an output beam that has the characteristics of laser light. The chapter closely examines various types of semiconductor lasers such as quantum cascade laser, quantum dot laser, transistor laser, interband cascade laser, quantum well laser, etc.

The semiconductor laser is very small in size and appearance. It is similar to a transistor and has the operation like LED but the output beam has the characteristics of laser light. The material which often used in semiconductor laser is the gallium Arsenide, therefore semiconductor laser is sometimes known as Gallium Arsenide Laser. It is also called Injection Laser. The semiconductor is made in unique manner for the semiconductor laser.

Design Features

In reality a semiconductor laser is simply a semiconductor diode, because its active medium is the junction of the forward biased P-N diode, shown as:

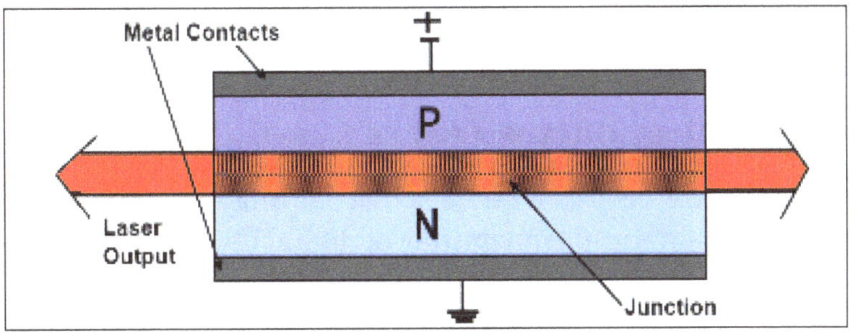

Here the metal contacts shown are used to connect the P-N material to the DC power supply. The junction shown is few micrometers thick. At the junction light is emitted when electrons or current pass from N to P type material. In other words, current is injected into the junction between N and P type materials. It is why we use to semiconductor laser the name of Injection Laser also.

Since we know that a minimum current density (similar to the gain threshold) is necessary for the occurrence of lasing. So when the minimum current density is reached then increasing the current density across the junction region will increase the output of the laser.

Unlike other lasers, semiconductor laser does not need mirrors to obtain the reflectivity needed to produce feedback mechanism. Reflection from the cleaved ends of the semiconductor chip is enough to produce lasing.

The reflectivity of the interface between the semiconductor material and air is approximately 36% which is enough to provide adequate feedback as well as serve as the output coupler. The beam although exist from both ends. If a beam is desired from one end of the laser only then opposite ends can be coated to reflect higher amounts of light.

The temperature has great effect on the output of the semiconductor laser. When the temperature increases then significant power losses occurs within the laser. That is why the semiconductor laser is sometimes cooled by liquid nitrogen or some other cooling system. However these lasers can be operated at room temperature if the losses are acceptable and current density is high enough.

Pulsing the current leads us to very much improved performance. Gallium Arsenide lasers are usually operated in pulsed mode, with duty cycle less than 1% because higher duty cycles cause an increased temperature, which greatly affect the output characteristics. The gallium arsenide laser produces light in near infrared spectrum ranging from 845nm to 905nm. The lasing medium of the semiconductor laser is short and rectangular. Therefore the output beam does not have the same dimension in both vertical and horizontal axis. Hence the beam profile has an unusual shape. The beam divergence of semiconductor lasers is much greater than most of the lasers, depending on temperature, therefore ranging from 125 to 400 milli radians.

In spite of the fact that semiconductor lasers do not produce a beam with characteristics as desired in other types of lasers, their small size, low power consumption and relatively high efficiency make them very useful device.

Application of Semiconductor Lasers

- The semiconductor laser can be pulsed at varying rate and pulse widths. Therefore this laser is a natural transmitter of digital data.
- Semiconductor laser is well suited for interface with fiber optic cables used in communication.

Advantages of Semiconductor Lasers

- Smaller size and appearance make them good choice for many applications.
- From cost point of view the semiconductor lasers are economical.
- Semiconductor lasers construction is very simple.
- No need of mirrors is in semiconductor lasers. Semiconductor lasers have high efficiency.
- The low power consumption is also its great advantage.

Disadvantages of Semiconductor Lasers

- Due to relatively low power production, these lasers are not suited to many typical laser applications.
- Semiconductor laser is greatly dependent on temperature. The temperature affects greatly the output of the laser.
- The lasing medium of semiconductor lasers is too short and rectangular so the output beam profile has an unusual shape.
- Beam divergence is much greater from 125 to 400 milli radians as compared to all other lasers. The cooling system requirement in some cases may be considered its disadvantage.

Laser Diode

A laser diode, (LD), injection laser diode (ILD), or diode laser is a semiconductor device similar to a light-emitting diode in which the laser beam is created at the diode's junction. Laser diodes can directly convert electrical energy into light. Driven by voltage, the doped p-n-transition allows for recombination of an electron with a hole. Due to the drop of the electron from a higher energy level to a lower one, radiation, in the form of an emitted photon is generated. This is spontaneous emission. Stimulated emission can be produced when the process is continued and further generate light with the same phase, coherence and wavelength.

The choice of the semiconductor material determines the wavelength of the emitted beam, which in today's laser diodes range from infra-red to the UV spectrum. Laser diodes are the most common type of lasers produced, with a wide range of uses that include fiber optic communications, barcode readers, laser pointers, CD/DVD/Blu-ray disc reading/recording, laser printing, laser scanning and light beam illumination.

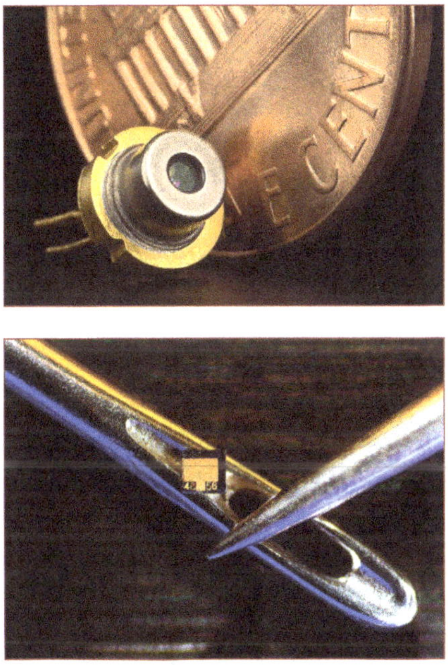

Top: a packaged laser diode shown with a penny for scale. Bottom: the laser diode chip is removed from the above package and placed on the eye of a needle for scale.

A laser diode with the case cut away. The laser diode chip is the small black chip at the front; a photodiode at the back is used to control output power.

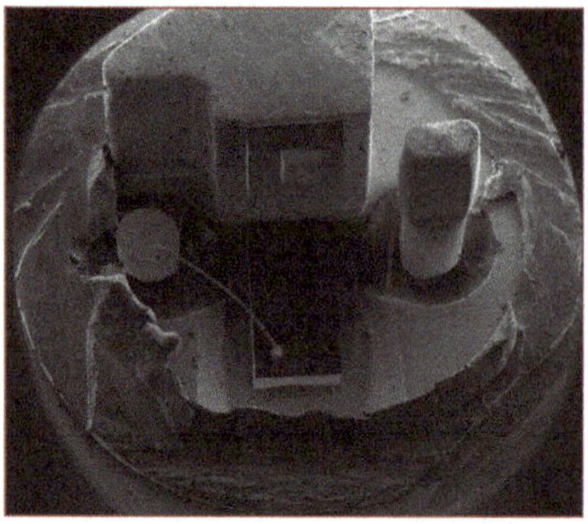

SEM (Scanning Electron Microscope) image of a commercial laser diode with its case and window cut away. The anode connection on the right has been accidentally broken by the case cut process.

Theory of Operation of Simple Diode

A laser diode is electrically a PIN diode. The active region of the laser diode is in the intrinsic (I) region, and the carriers (electrons and holes) are pumped into that region from the N and P regions respectively. While initial diode laser research was conducted on simple P-N diodes, all modern lasers use the double-hetero-structure implementation, where the carriers and the photons are confined in order to maximize their chances for recombination and light generation. Unlike a regular diode, the goal for a laser diode is to recombine all carriers in the I region, and produce light. Thus, laser diodes are fabricated using direct band-gap semiconductors. The laser diode epitaxial structure is grown using one of the crystal growth techniques, usually starting from an N doped substrate, and growing the I doped active layer, followed by the P doped cladding, and a contact layer. The active layer most often consists of quantum wells, which provide lower threshold current and higher efficiency.

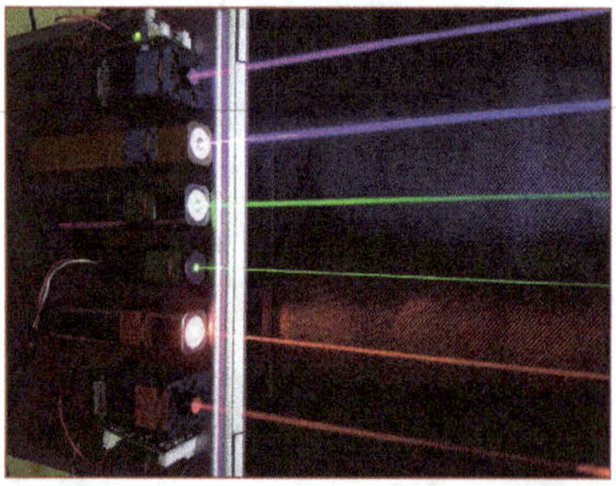

Semi-conductor lasers (660 nm, 635 nm, 532 nm, 520 nm, 445 nm, 405.

Electrical and Optical Pumping

Laser diodes form a subset of the larger classification of semiconductor *p-n* junction diodes. Forward electrical bias across the laser diode causes the two species of charge carrier – holes and electrons – to be "injected" from opposite sides of the *p-n* junction into the depletion region. Holes are injected from the *p*-doped, and electrons from the *n*-doped, semiconductor. (A depletion region, devoid of any charge carriers, forms as a result of the difference in electrical potential between *n*- and *p*-type semiconductors wherever they are in physical contact.) Due to the use of charge injection in powering most diode lasers, this class of lasers is sometimes termed "injection lasers," or "injection laser diode" (ILD). As diode lasers are semiconductor devices, they may also be classified as semiconductor lasers. Either designation distinguishes diode lasers from solid-state lasers.

Another method of powering some diode lasers is the use of optical pumping. Optically pumped semiconductor lasers (OPSL) use a III-V semiconductor chip as the gain medium, and another laser (often another diode laser) as the pump source. OPSL offer several advantages over ILDs, particularly in wavelength selection and lack of interference from internal electrode structures. A further advantage of OPSLs is invariance of the beam parameters - divergence, shape, and pointing - as pump power (and hence output power) is varied, even over a 10:1 output power ratio.

Generation of Spontaneous Emission

When an electron and a hole are present in the same region, they may recombine or "annihilate" producing a spontaneous emission — i.e., the electron may re-occupy the energy state of the hole, emitting a photon with energy equal to the difference between the electron's original state and hole's state. (In a conventional semiconductor junction diode, the energy released from the recombination of electrons and holes is carried away as phonons, i.e., lattice vibrations, rather than as photons.) Spontaneous emission below the lasing threshold produces similar properties to an LED. Spontaneous emission is necessary to initiate laser oscillation, but it is one among several sources of inefficiency once the laser is oscillating.

Direct and Indirect Bandgap Semiconductors

The difference between the photon-emitting semiconductor laser and a conventional phonon-emitting (non-light-emitting) semiconductor junction diode lies in the type of semiconductor used, one whose physical and atomic structure confers the possibility for photon emission. These photon-emitting semiconductors are the so-called "direct bandgap" semiconductors. The properties of silicon and germanium, which are single-element semiconductors, have bandgaps that do not align in the way needed to allow photon emission and are not considered "direct." Other materials, the so-called compound semiconductors, have virtually identical crystalline structures as silicon or germanium but use alternating arrangements of two different atomic species in a checkerboard-like pattern to break the symmetry. The transition between the materials in the alternating pattern creates the critical "direct bandgap" property. Gallium arsenide, indium phosphide, gallium antimonide, and gallium nitride are all examples of compound semiconductor materials that can be used to create junction diodes that emit light.

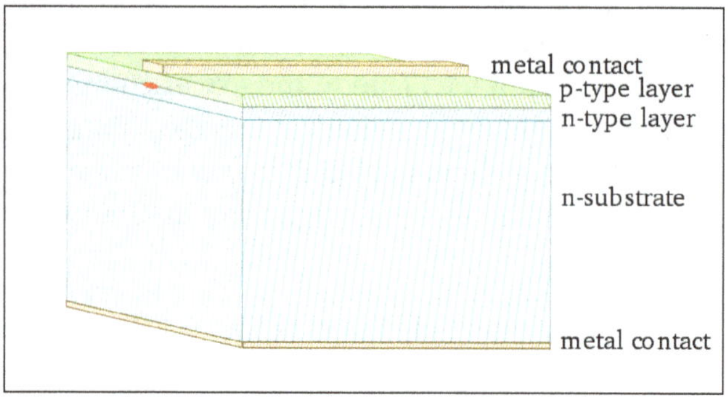

Diagram of a simple laser diode, such as shown above; not to scale.

A simple and low power metal enclosed laser diode.

Generation of Stimulated Emission

In the absence of stimulated emission (e.g., lasing) conditions, electrons and holes may coexist in proximity to one another, without recombining, for a certain time, termed the "upper-state lifetime" or "recombination time" (about a nanosecond for typical diode laser materials), before they recombine. A nearby photon with energy equal to the recombination energy can cause recombination by stimulated emission. This generates another photon of the same frequency, polarization, and phase, travelling in the same direction as the first photon. This means that stimulated emission will cause gain in an optical wave (of the correct wavelength) in the injection region, and the gain increases as the number of electrons and holes injected across the junction increases. The spontaneous and stimulated emission processes are vastly more efficient in direct bandgap semiconductors than in indirect bandgap semiconductors; therefore silicon is not a common material for laser diodes.

Optical Cavity and Laser Modes

As in other lasers, the gain region is surrounded with an optical cavity to form a laser. In the simplest form of laser diode, an optical waveguide is made on that crystal's surface, such that the light is confined to a relatively narrow line. The two ends of the crystal are cleaved to form perfectly smooth, parallel edges, forming a Fabry–Pérot resonator. Photons emitted into a mode of the waveguide will travel along the waveguide and be reflected several times from each end face before they exit. As a light wave passes through the cavity, it is amplified by stimulated emission, but light is also lost due to absorption and by incomplete reflection from the end facets. Finally, if there is more amplification than loss, the diode begins to "lase".

Some important properties of laser diodes are determined by the geometry of the optical cavity. Generally, the light is contained within a very thin layer, and the structure supports only a single optical mode in the direction perpendicular to the layers. In the transverse direction, if the waveguide is wide compared to the wavelength of light, then the waveguide can support multiple transverse optical modes, and the laser is known as "multi-mode". These transversely multi-mode lasers are adequate in cases where one needs a very large amount of power, but not a small diffraction-limited beam; for example in printing, activating chemicals, or pumping other types of lasers.

In applications where a small focused beam is needed, the waveguide must be made narrow, on the order of the optical wavelength. This way, only a single transverse mode is supported and one ends up with a diffraction-limited beam. Such single spatial mode devices are used for optical storage, laser pointers, and fiber optics. Note that these lasers may still support multiple longitudinal modes, and thus can lase at multiple wavelengths simultaneously. The wavelength emitted is a function of the band-gap of the semiconductor material and the modes of the optical cavity. In general, the maximum gain will occur for photons with energy slightly above the band-gap energy, and the modes nearest the peak of the gain curve will lase most strongly. The width of the gain curve will determine the number of additional "side modes" that may also lase, depending on the operating conditions. Single spatial mode lasers that can support multiple longitudinal modes are called Fabry Perot (FP) lasers. An FP laser will lase at multiple cavity modes within the gain bandwidth of the lasing medium. The number of lasing modes in an FP laser is usually unstable, and can fluctuate due to changes in current or temperature.

Single spatial mode diode lasers can be designed so as to operate on a single longitudinal mode. These single frequency diode lasers exhibit a high degree of stability, and are used in spectroscopy and metrology, and as frequency references. Single frequency diode lasers are classed as either distributed feedback (DFB) lasers or distributed Bragg reflector (DBR) lasers.

Formation of Laser Beam

Due to diffraction, the beam diverges (expands) rapidly after leaving the chip, typically at 30 degrees vertically by 10 degrees laterally. A lens must be used in order to form a collimated beam like that produced by a laser pointer. If a circular beam is required, cylindrical lenses and other optics are used. For single spatial mode lasers, using symmetrical lenses, the collimated beam ends up being elliptical in shape, due to the difference in the vertical and lateral divergences. This is easily observable with a red laser pointer.

Types

The simple laser diode structure, described above, is extremely inefficient. Such devices require so much power that they can only achieve pulsed operation without damage. Although historically important and easy to explain, such devices are not practical.

Double Heterostructure Lasers

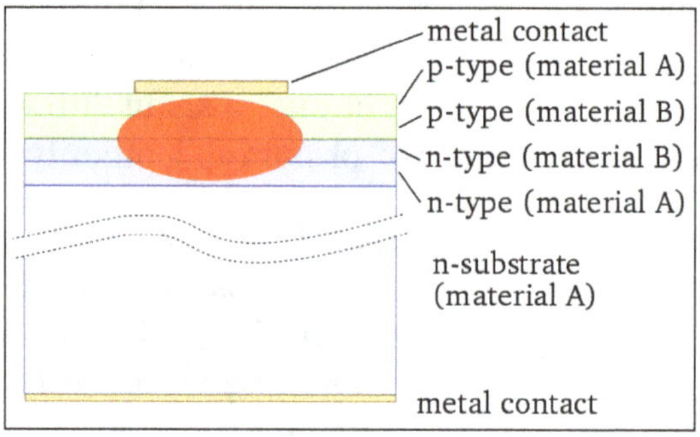

Diagram of front view of a double heterostructure laser diode.

In these devices, a layer of low bandgap material is sandwiched between two high bandgap layers. One commonly-used pair of materials is gallium arsenide (GaAs) with aluminium gallium arsenide ($Al_xGa_{(1-x)}As$). Each of the junctions between different bandgap materials is called a *heterostructure*, hence the name "double heterostructure laser" or *DH* laser. The kind of laser diode described in the first part of the article may be referred to as a *homojunction* laser, for contrast with these more popular devices.

The advantage of a DH laser is that the region where free electrons and holes exist simultaneously—the active region—is confined to the thin middle layer. This means that many more of the electron-hole pairs can contribute to amplification—not so many are left out in the poorly amplifying periphery. In addition, light is reflected within the heterojunction; hence, the light is confined to the region where the amplification takes place.

Quantum Well Lasers

If the middle layer is made thin enough, it acts as a quantum well. This means that the vertical variation of the electron's wavefunction, and thus a component of its energy, is quantized. The efficiency of a quantum well laser is greater than that of a bulk laser because the density of states function of electrons in the quantum well system has an abrupt edge that concentrates electrons in energy states that contribute to laser action.

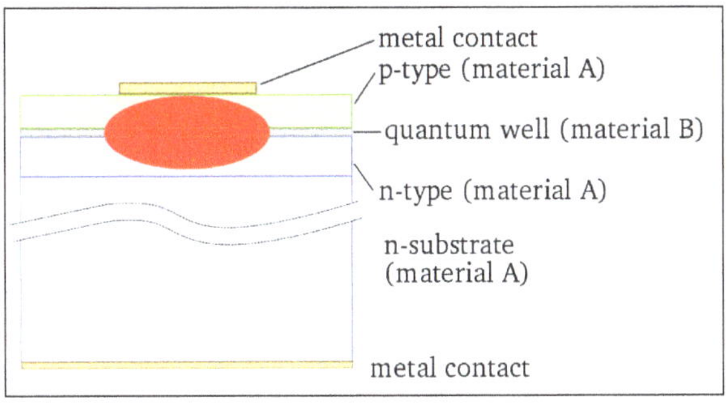

Diagram of front view of a simple quantum well laser diode.

Lasers containing more than one quantum well layer are known as *multiple quantum well* lasers. Multiple quantum wells improve the overlap of the gain region with the optical waveguide mode.

Further improvements in the laser efficiency have also been demonstrated by reducing the quantum well layer to a quantum wire or to a "sea" of quantum dots.

Quantum Cascade Lasers

In a quantum cascade laser, the difference between quantum well energy levels is used for the laser transition instead of the bandgap. This enables laser action at relatively long wavelengths, which can be tuned simply by altering the thickness of the layer. They are heterojunction lasers.

Interband Cascade Lasers

A Interband cascade laser (ICL) is a type of laser diode that can produce coherent radiation over a large part of the mid-infrared region of the electromagnetic spectrum.

Separate Confinement Heterostructure Lasers

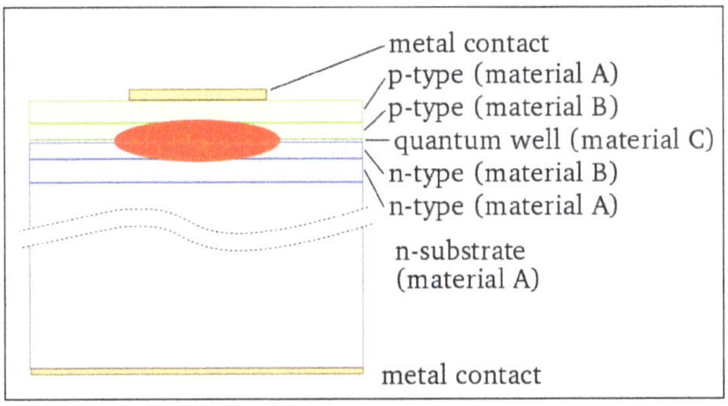

Diagram of front view of a separate confinement heterostructure quantum well laser diode.

The problem with the simple quantum well diode described above is that the thin layer is simply too small to effectively confine the light. To compensate, another two layers are added on, outside the first three. These layers have a lower refractive index than the centre layers, and hence confine the light effectively. Such a design is called a separate confinement heterostructure (SCH) laser diode.

Almost all commercial laser diodes since the 1990s have been SCH quantum well diodes.

Distributed Bragg Re lector Lasers

A distributed Bragg reflector laser (DBR) is a type of single frequency laser diode. It is characterized by an optical cavity consisting of an electrically or optically pumped gain region between two mirrors to provide feedback. One of the mirrors is a broadband reflector and the other mirror is wavelength selective so that gain is favored on a single longitudinal mode, resulting in lasing at a single resonant frequency. The broadband mirror is usually coated with a low reflectivity coating to allow emission. The wavelength selective mirror is a periodically structured diffraction grating with high reflectivity. The diffraction grating is within a non-pumped, or passive region of the cavity. A DBR laser is a monolithic single chip device with the grating etched into the semiconductor. DBR lasers can be edge emitting lasers or VCSELs. Alternative hybrid architectures that share the same topology include extended cavity diode lasers and volume Bragg grating lasers, but these are not properly called DBR lasers.

Distributed Feedback Lasers

A distributed feedback laser (DFB) is a type of single frequency laser diode. DFBs are the most common transmitter type in DWDM-systems. To stabilize the lasing wavelength, a diffraction grating is etched close to the p-n junction of the diode. This grating acts like an optical filter, causing a single wavelength to be fed back to the gain region and lase. Since the grating provides the feedback that is required for lasing, reflection from the facets is not required. Thus, at least one facet of a DFB is anti-reflection coated. The DFB laser has a stable wavelength that is set during manufacturing by the pitch of the grating, and can only be tuned slightly with temperature. DFB lasers are widely used in optical communication applications where a precise and stable wavelength is critical.

The threshold current of this DFB laser, based on its static characteristic, is around 11 mA. The appropriate bias current in a linear regime could be taken in the middle of the static characteristic (50 mA). Several techniques have been proposed in order to enhance the single-mode operation in these kinds of lasers by inserting a onephase-shift (1PS) or multiple-phase-shift (MPS) in the uniform Bragg grating. However, multiple-phase-shift DFB lasers represent the optimal solution because they have the combination of higher side-mode suppression ratio and reduced spatial hole-burning.

Vertical-cavity Surface-emitting Laser

Vertical-cavity surface-emitting lasers (VCSELs) have the optical cavity axis along the direction of current flow rather than perpendicular to the current flow as in conventional laser diodes. The active region length is very short compared with the lateral dimensions so that the radiation emerges from the surface of the cavity rather than from its edge as shown in the figure. The reflectors at the ends of the cavity are dielectric mirrors made from alternating high and low refractive index quarter-wave thick multilayer.

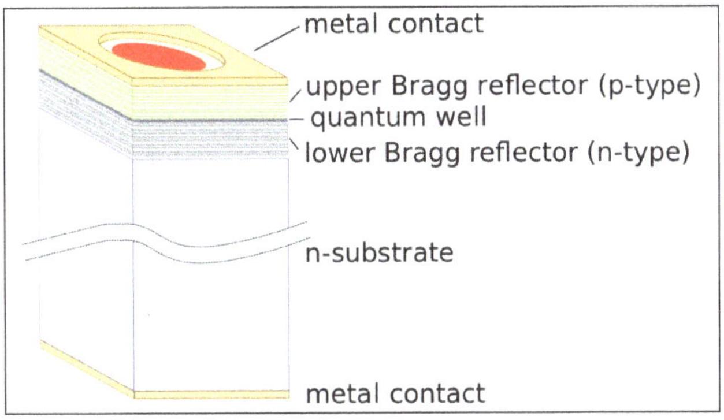

Diagram of a simple VCSEL structure.

Such dielectric mirrors provide a high degree of wavelength-selective reflectance at the required free surface wavelength λ if the thicknesses of alternating layers d_1 and d_2 with refractive indices n_1 and n_2 are such that $n_1d_1 + n_2d_2 = \lambda/2$ which then leads to the constructive interference of all partially reflected waves at the interfaces. But there is a disadvantage: because of the high mirror reflectivities, VCSELs have lower output powers when compared to edge-emitting lasers.

There are several advantages to producing VCSELs when compared with the production process of edge-emitting lasers. Edge-emitters cannot be tested until the end of the production process. If the edge-emitter does not work, whether due to bad contacts or poor material growth quality, the production time and the processing materials have been wasted.

Additionally, because VCSELs emit the beam perpendicular to the active region of the laser as opposed to parallel as with an edge emitter, tens of thousands of VCSELs can be processed simultaneously on a three-inch gallium arsenide wafer. Furthermore, even though the VCSEL production process is more labor- and material-intensive, the yield can be controlled to a more predictable outcome. However, they normally show a lower power output level.

Vertical-external-cavity Surface-emitting-laser

Vertical external-cavity surface-emitting lasers, or VECSELs, are similar to VCSELs. In VCSELs, the mirrors are typically grown epitaxially as part of the diode structure, or grown separately and bonded directly to the semiconductor containing the active region. VECSELs are distinguished by a construction in which one of the two mirrors is external to the diode structure. As a result, the cavity includes a free-space region. A typical distance from the diode to the external mirror would be 1 cm.

One of the most interesting features of any VECSEL is the small thickness of the semiconductor gain region in the direction of propagation, less than 100 nm. In contrast, a conventional in-plane semiconductor laser entails light propagation over distances of from 250 μm upward to 2 mm or longer. The significance of the short propagation distance is that it causes the effect of "antiguiding" nonlinearities in the diode laser gain region to be minimized. The result is a large-cross-section single-mode optical beam which is not attainable from in-plane ("edge-emitting") diode lasers.

Several workers demonstrated optically pumped VECSELs, and they continue to be developed for many applications including high power sources for use in industrial machining (cutting, punching, etc.) because of their unusually high power and efficiency when pumped by multi-mode diode laser bars. However, because of their lack of p-n junction, optically-pumped VECSELs are not considered "diode lasers", and are classified as semiconductor lasers.

Electrically pumped VECSELs have also been demonstrated. Applications for electrically pumped VECSELs include projection displays, served by frequency doubling of near-IR VECSEL emitters to produce blue and green light.

External-cavity Diode Lasers

External-cavity diode lasers are tunable lasers which use mainly double heterostructures diodes of the $Al_xGa_{(1-x)}As$ type. The first external-cavity diode lasers used intracavity etalons and simple tuning Littrow gratings. Other designs include gratings in grazing-incidence configuration and multiple-prism grating configurations.

Failure Mechanisms

Laser diodes have the same reliability and failure issues as light emitting diodes. In addition they are subject to *catastrophic optical damage* (COD) when operated at higher power.

Many of the advances in reliability of diode lasers in the last 20 years remain proprietary to their developers. The reliability of a laser diode can make or break a product line. Moreover, *reverse engineering* is not always able to reveal the differences between more-reliable and less-reliable diode laser products.

At the edge of a diode laser, where light is emitted, a mirror is traditionally formed by cleaving the semiconductor wafer to form a specularly reflecting plane. This approach is facilitated by the weakness of the crystallographic plane in III-V semiconductor crystals (such as GaAs, InP, GaSb, etc.) compared to other planes. A scratch made at the edge of the wafer and a slight bending force causes a nearly atomically perfect mirror-like cleavage plane to form and propagate in a straight line across the wafer.

But it so happens that the atomic states at the cleavage plane are altered (compared to their bulk properties within the crystal) by the termination of the perfectly periodic lattice at that plane. Surface states at the cleaved plane have energy levels within the (otherwise forbidden) bandgap of the semiconductor.

Essentially, as a result, when light propagates through the cleavage plane and transits to free space from within the semiconductor crystal, a fraction of the light energy is absorbed by the surface states where it is converted to heat by phonon-electron interactions. This heats the cleaved mirror. In addition, the mirror may heat simply because the edge of the diode laser—which is electrically pumped—is in less-than-perfect contact with the mount that provides a path for heat removal. The heating of the mirror causes the bandgap of the semiconductor to shrink in the warmer areas. The bandgap shrinkage brings more electronic band-to-band transitions into alignment with the photon energy causing yet more absorption. This is thermal runaway, a form of positive feedback, and the result can be melting of the facet, known as *catastrophic optical damage*, or COD.

In the 1970s, this problem, which is particularly nettlesome for GaAs-based lasers emitting between 0.630 μm and 1 μm wavelengths (less so for InP-based lasers used for long-haul telecommunications which emit between 1.3 μm and 2 μm), was identified. Michael Ettenberg, a researcher and later Vice President at RCA Laboratories' David Sarnoff Research Center in Princeton, New Jersey, devised a solution. A thin layer of aluminum oxide was deposited on the facet. If the aluminum oxide thickness is chosen correctly, it functions as an anti-reflective coating, reducing reflection at the surface. This alleviated the heating and COD at the facet.

Since then, various other refinements have been employed. One approach is to create a so-called non-absorbing mirror (NAM) such that the final 10 μm or so before the light emits from the cleaved facet are rendered non-absorbing at the wavelength of interest.

Reliability of high-power diode laser pump bars (used to pump solid-state lasers) remains a difficult problem in a variety of applications, in spite of these proprietary advances. Indeed, the physics of diode laser failure is still being worked out and research on this subject remains active, if proprietary.

Extension of the lifetime of laser diodes is critical to their continued adaptation to a wide variety of applications.

Applications

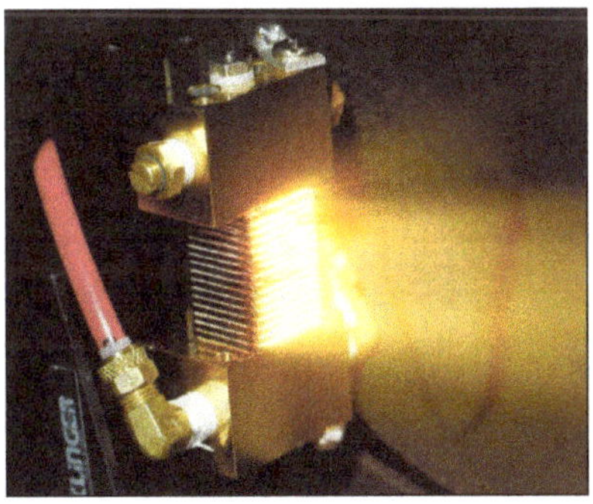

Laser diodes can be arrayed to produce very high power outputs, continuous wave or pulsed. Such arrays may be used to efficiently pump solid-state lasers for high average power drilling, burning or for inertial confinement fusion.

Laser diodes are numerically the most common laser type, with 2004 sales of approximately 733 million units, as compared to 131,000 of other types of lasers.

Telecommunications, Scanning and Spectrometry

Laser diodes find wide use in telecommunication as easily modulated and easily coupled light sources for fiber optics communication. They are used in various measuring instruments, such as rangefinders. Another common use is in barcode readers. Visible lasers, typically red but later also green, are common as laser pointers. Both low and high-power diodes are used extensively in the printing industry both as light sources for scanning (input) of images and for very high-speed and high-resolution printing plate (output) manufacturing. Infrared and red laser diodes are common in CD players, CD-ROMs and DVD technology. Violet lasers are used in HD DVD and Blu-ray technology. Diode lasers have also found many applications in laser absorption spectrometry (LAS) for high-speed, low-cost assessment or monitoring of the concentration of various species in gas phase. High-power laser diodes are used in industrial applications such as heat treating, cladding, seam welding and for pumping other lasers, such as diode-pumped solid-state lasers.

Uses of laser diodes can be categorized in various ways. Most applications could be served by larger solid-state lasers or optical parametric oscillators, but the low cost of mass-produced diode lasers makes them essential for mass-market applications. Diode lasers can be used in a great many fields; since light has many different properties (power, wavelength, spectral and beam quality, polarization, etc.) it is useful to classify applications by these basic properties.

Many applications of diode lasers primarily make use of the "directed energy" property of an optical beam. In this category, one might include the laser printers, barcode readers, image scanning, illuminators, designators, optical data recording, combustion ignition, laser surgery, industrial sorting, industrial machining, and directed energy weaponry. Some of these applications are well-established while others are emerging.

Medical Uses

Laser medicine: medicine and especially dentistry have found many new uses for diode lasers. The shrinking size and cost of the units and their increasing user friendliness makes them very attractive to clinicians for minor soft tissue procedures. Diode wavelengths range from 810 to 1,100 nm, are poorly absorbed by soft tissue, and are not used for cutting or ablation. Soft tissue is not cut by the laser's beam, but is instead cut by contact with a hot charred glass tip. The laser's irradiation is highly absorbed at the distal end of the tip and heats it up to 500 °C to 900 °C. Because the tip is so hot, it can be used to cut soft-tissue and can cause hemostasis through cauterization and carbonization. Diode lasers when used on soft tissue can cause extensive collateral thermal damage to surrounding tissue.

As laser beam light is inherently coherent, certain applications utilize the coherence of laser diodes. These include interferometric distance measurement, holography, coherent communications, and coherent control of chemical reactions.

Laser diodes are used for their "narrow spectral" properties in the areas of range-finding, telecommunications, infra-red countermeasures, spectroscopic sensing, generation of radio-frequency or terahertz waves, atomic clock state preparation, quantum key cryptography, frequency doubling and conversion, water purification (in the UV), and photodynamic therapy (where a particular wavelength of light would cause a substance such as porphyrin to become chemically active as an anti-cancer agent only where the tissue is illuminated by light).

Laser diodes are used for their ability to generate ultra-short pulses of light by the technique known as "mode-locking." Areas of use include clock distribution for high-performance integrated circuits, high-peak-power sources for laser-induced breakdown spectroscopy sensing, arbitrary waveform generation for radio-frequency waves, photonic sampling for analog-to-digital conversion, and optical code-division-multiple-access systems for secure communication.

Common Wavelengths and Uses

Visible light405 nm – InGaN blue-violet laser, in Blu-ray Disc and HD DVD drives:

- 445–465 nm – InGaN blue laser multimode diode recently introduced (2010) for use in mercury-free high-brightness data projectors.
- 510–525 nm – InGaN Green diodes recently (2010) developed by Nichia and OSRAM for laser projectors.
- 635 nm – AlGaInP better red laser pointers, same power subjectively twice as bright as 650 nm.
- 650–660 nm – GaInP/AlGaInP CD and DVD drives, cheap red laser pointers.
- 670 nm – AlGaInP bar code readers, first diode laser pointers (now obsolete, replaced by brighter 650 nm and 671 nm DPSS).

Infrared

- 760 nm – AlGaInP gas sensing: O_2.
- 785 nm – GaAlAs Compact Disc drives.
- 808 nm – GaAlAs pumps in DPSS Nd:YAG lasers (e.g., in green laser pointers or as arrays in higher-powered lasers).
- 848 nm – laser mice.

- 980 nm – InGaAs pump for optical amplifiers, for Yb:YAG DPSS lasers.
- 1,064 nm – AlGaAs fiber-optic communication, DPSS laser pump frequency.
- 1,310 nm – InGaAsP, InGaAsN fiber-optic communication.
- 1,480 nm – InGaAsP pump for optical amplifiers.
- 1,512 nm – InGaAsP gas sensing: NH_3.
- 1,550 nm – InGaAsP, InGaAsNSb fiber-optic communication.
- 1,625 nm – InGaAsP fiber-optic communication, service channel.
- 1,654 nm – InGaAsP gas sensing: CH_4.
- 1,877 nm – GaInAsSb gas sensing: H_2O.
- 2,004 nm – GaInAsSb gas sensing: CO_2.
- 2,330 nm – GaInAsSb gas sensing: CO.
- 2,680 nm – GaInAsSb gas sensing: CO_2.
- 3,030 nm – GaInAsSb gas sensing: C_2H_2.
- 3,330 nm – GaInAsSb gas sensing: CH_4.

Distributed Feedback Laser

A distributed feedback laser (DFB) is a type of laser diode, quantum cascade laser or optical fiber laser where the active region of the device contains a periodically structured element or diffraction grating. The structure builds a one-dimensional interference grating (Bragg scattering) and the grating provides optical feedback for the laser. This longitudinal diffraction grating has periodic changes in refractive index that cause reflection back into the cavity. The periodic change can be either in the real part of the refractive index, or in the imaginary part (gain or absorption). The strongest grating operates in the first order - where the periodicity is one-half wave, and the light is reflected backwards. DFB lasers tend to be much more stable than Fabry-Perot or DBR lasers and are used frequently when clean single mode operation is needed, especially in high speed fiber optic telecommunications. Semiconductor DFB lasers in the lowest loss window of optical fibers at about 1.55um wavelength, amplified by Erbium-doped fiber amplifiers (EDFAs), dominate the long distance communication market, while DFB lasers in the lowest dispersion window at 1.3um are used at shorter distances.

The simplest kind of a laser is a Fabry-Perot laser, where there are two broad-band reflectors at the two ends of the lasing optical cavity. The light bounces back and forth between these two mirrors and forms longitudinal modes or standing waves. The back reflector is generally high reflectivity, and the front mirror is lower reflectivity. The light then leaks out of the front mirror and forms the output of the laser diode. Since the mirrors are generally broad-band and reflect many wavelengths, the laser supports multiple longitudinal modes, or standing waves, simultaneously and lases multimode, or easily jumps between longitudinal modes. If the temperature of a semiconductor Fabry-Perot laser changes, the wavelengths that are amplified by the lasing medium vary rapidly. At the same time, the longitudinal modes of the laser also vary, as the refractive index is also a function of temperature. This causes the spectrum to be unstable and highly temperature dependent. At the important wavelengths of 1.55um and 1.3um, the peak gain typically moves about 0.4nm to the longer wavelengths as the temperature increases, while the longitudinal modes shift about 0.1nm to the longer wavelengths.

If one or both of these end mirrors are replaced with a diffraction grating, the structure is then known as a DBR laser (Distributed Bragg Reflector). These longitudinal diffraction grating mirrors reflect the light back in the cavity, very much like a multi-layer mirror coating. The diffraction grating mirrors tend be reflect a narrower band of wavelengths than normal end mirrors, and this limits the number of standing waves that can be supported by the gain in the cavity. So DBR lasers tend to be more spectrally stable than Fabry-Perot lasers with broadband coatings. Nevertheless, as the temperature or current changes in the laser, the device can "mode-hop" jumping from one standing wave to another. The overall shifts with temperature are however lower with DBR lasers as the mirrors determine which longitudinal modes lase, and they shift with the refractive index and not the peak gain.

In a DFB laser, the grating and the reflection is generally continuous along the cavity, instead of just being at the two ends. This changes the modal behavior considerably and makes the laser more stable. There are various designs of DFB lasers, each with slightly different properties.

If the grating is periodic and continuous, and the ends of the laser are anti-reflection (AR/AR) coated, so there is no feedback other than the grating itself, then such a structure supports two longitudinal (degenerate) modes and almost always lases at two wavelengths. Obviously a two-moded laser is generally not desirable. So there are various ways of breaking this "degeneracy".

The first is by inducing a quarter-wave shift in the cavity. This phase-shift acts a like a "defect" and creates a resonance in the center of the reflectivity bandwidth or "stop-band." The laser then lases at this resonance and is extremely stable. As the temperature and current changes, the grating and the cavity shift together at the lower rate of the refractive index change, and there are no modehops. However, light is emitted from both sides of the lasers, and generally the light from one side is wasted. Furthermore, creating an exact quarter-wave shift can be technologically difficult to achieve, and often requires directly-written electron-beam lithography.

Often, rather than a single quarter-wave phase shift at the center of the cavity, multiple smaller shifts distributed in the cavity at different locations that spread out the mode longitudinally and give higher output power.

An alternate way of breaking this degeneracy is by coating the back end of the laser to a high reflectivity (HR). The exact position of this end reflector cannot be accurately controlled, and so one obtains a random phase shift between the grating and the exact position of the end mirror. Sometimes this leads to a perfect phase shift, where effectively a quarter-wave phase shifted DFB is reflected on itself. In this case all the light exits the front facet and one obtains a very stable laser. At other times, however, the phase shift between the grating and the high-reflector back mirror is not optimal, and one ends up with a two-moded lasers again. Additionally, the phase of the cleave affects the wavelength, and thus controlling the output wavelength of a batch of lasers in manufacturing can be a challenge. Thus the HR/AR DFB lasers tend to be low yield and have to be screened before use. There are various combinations of coatings and phase shifts that can be optimized for power and yield, and generally each manufacturer has their own technique to optimize performance and yield.

To encode data on a DFB laser for fiber optic communications, generally the electric drive current is varied to modulate the intensity of the light. These DMLs (Directly modulated lasers) are the simplest kinds and are found in various fiber optic systems. The disadvantage of directly modulating a laser is that there are associated frequency shifts together with the intensity shifts (laser chirp). These frequency shifts, together with dispersion in the fiber, cause the signal to degrade after some distance, limiting the bandwidth and the range. An alternate structure is an electro-absorption modulated laser (EML) that runs the laser continuously and has a separate section integrated in front that either absorbs or transmits the light - very much like an optical shutter. These EMLs can operate at higher speeds and have much lower chirp. In very high performance coherent optical communication systems, the DFB laser is run continuously and is followed by a phase modulator. On the receiving end, a local oscillator DFB interferes with the received signal and decodes the modulation.

An alternative approach is a phase-shifted DFB laser. In this case both facets are anti-reflection coated and there is a phase shift in the cavity. Such devices have much better reproducibility in wavelength and theoretically all lase in single mode.

In DFB fibre lasers the Bragg grating (which in this case forms also the cavity of the laser) has a phase-shift centered in the reflection band akin to a single very narrow transmission notch of a Fabry–Pérot interferometer. When configured properly, these lasers operate on a single longitudinal mode with coherence lengths in excess of tens of kilometres, essentially limited by the temporal noise induced by the self-heterodyne coherence detection technique used to measure the coherence. These DFB fibre lasers are often used in sensing applications where extreme narrow line width is required.

Vertical-cavity Surface-emitting Laser

The vertical-cavity surface-emitting laser, or VCSEL is a type of semiconductor laser diode with laser beam emission perpendicular from the top surface, contrary to conventional edge-emitting semiconductor lasers (also in-plane lasers) which emit from surfaces formed by cleaving the individual chip out of a wafer. VCSELs are used in various laser products, including computer mice, fiber optic communications, laser printers, Face ID, and smartglasses.

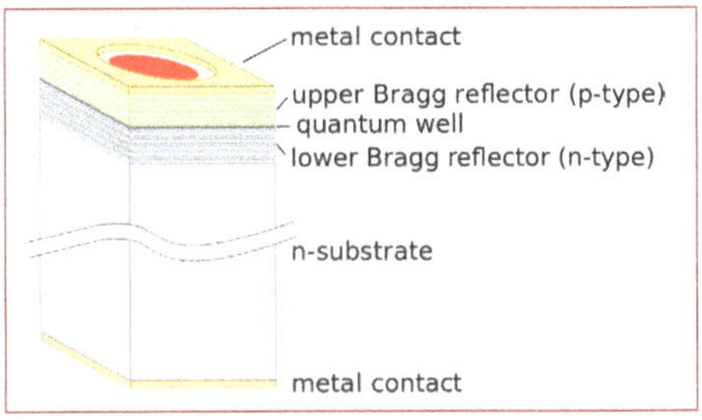

Diagram of a simple VCSEL structure.

Production Advantages

There are several advantages to producing VCSELs, in contrast to the production process of edge-emitting lasers. Edge-emitters cannot be tested until the end of the production process. If the edge-emitter does not function properly, whether due to bad contacts or poor material growth quality, the production time and the processing materials have been wasted. VCSELs however, can be tested at several stages throughout the process to check for material quality and processing issues. For instance, if the vias, the electrical connections between layers of a circuit, have not been completely cleared of dielectric material during the etch, an interim testing process will flag that the top metal layer is not making contact to the initial metal layer. Additionally, because VCSELs emit the beam perpendicular to the active region of the laser as opposed to parallel as with an edge emitter, tens of thousands of VCSELs can be processed simultaneously on a three-inch gallium arsenide wafer. Furthermore, even though the VCSEL production process is more labor and material intensive, the yield can be controlled to a more predictable outcome.

Structure

The laser resonator consists of two distributed Bragg reflector (DBR) mirrors parallel to the wafer surface with an active region consisting of one or more quantum wells for the laser light generation in between. The planar DBR-mirrors consist of layers with alternating high and low refractive indices. Each layer has a thickness of a quarter of the laser wavelength in the material, yielding intensity reflectivities above 99%. High reflectivity mirrors are required in VCSELs to balance the short axial length of the gain region.

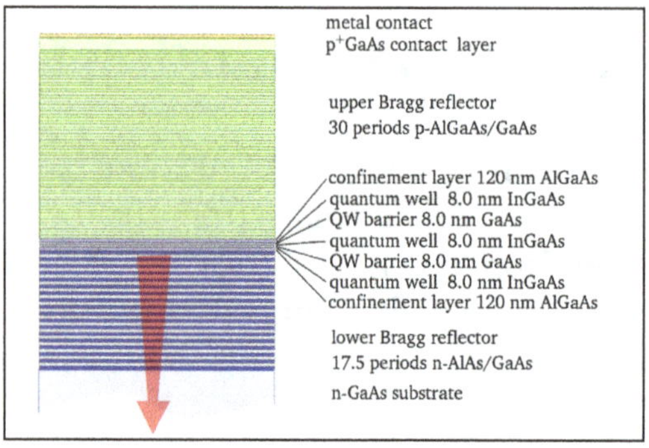

A realistic VCSEL device structure. This is a bottom-emitting multiple-quantum-well VCSEL.

In common VCSELs the upper and lower mirrors are doped as p-type and n-type materials, forming a diode junction. In more complex structures, the p-type and n-type regions may be embedded between the mirrors, requiring a more complex semiconductor process to make electrical contact to the active region, but eliminating electrical power loss in the DBR structure.

In laboratory investigation of VCSELs using new material systems, the active region may be *pumped* by an external light source with a shorter wavelength, usually another laser. This allows a VCSEL to be demonstrated without the additional problem of achieving good electrical performance; however such devices are not practical for most applications.

VCSELs for wavelengths from 650 nm to 1300 nm are typically based on gallium arsenide (GaAs) wafers with DBRs formed from GaAs and aluminium gallium arsenide ($Al_xGa_{(1-x)}As$). The GaAs–AlGaAs system is favored for constructing VCSELs because the lattice constant of the material does not vary strongly as the composition is changed, permitting multiple "lattice-matched" epitaxial layers to be grown on a GaAs substrate. However, the refractive index of AlGaAs does vary relatively strongly as the Al fraction is increased, minimizing the number of layers required to form an efficient Bragg mirror compared to other candidate material systems. Furthermore, at high aluminium concentrations, an oxide can be formed from AlGaAs, and this oxide can be used to restrict the current in a VCSEL, enabling very low threshold currents.

The main methods of restricting the current in a VCSEL are characterized by two types: ion-implanted VCSELs and oxide VCSELs. In the early 1990s, telecommunications companies tended to favor ion-implanted VCSELs. Ions, (often hydrogen ions, H^+), were implanted into the VCSEL structure everywhere except the aperture of the VCSEL, destroying the lattice structure around the aperture, thus inhibiting the current. In the mid to late 1990s, companies moved towards the technology of oxide VCSELs. The current is confined in an oxide VCSEL by oxidizing the material around the aperture of the VCSEL. A high content aluminium layer that is grown within the VCSEL structure is the layer that is oxidized. Oxide VCSELs also often employ the ion implant production step. As a result, in the oxide VCSEL, the current path is confined by the ion implant and the oxide aperture.

The initial acceptance of oxide VCSELs was plagued with concern about the apertures "popping off" due to the strain and defects of the oxidation layer. However, after much testing, the reliability of the structure has proven to be robust. As stated in one study by Hewlett Packard on oxide VCSELs, "The stress results show that the activation energy and the wearout lifetime of oxide VCSEL are similar to that of implant VCSEL emitting the same amount of output power." A production concern also plagued the industry when moving the oxide VCSELs from research and development to production mode. The oxidation rate of the oxide layer was highly dependent on the aluminium content. Any slight variation in aluminium would change the oxidation rate sometimes resulting in apertures that were either too big or too small to meet the specification standards.

Longer wavelength devices, from 1300 nm to 2000 nm, have been demonstrated with at least the active region made of indium phosphide. VCSELs at even higher wavelengths are experimental and usually optically pumped. 1310 nm VCSELs are desirable as the dispersion of silica-based optical fiber is minimal in this wavelength range.

Special Forms

Multiple Active Region Devices

Allows for differential quantum efficiency values in excess of 100% through carrier recycling.

VCSELs with Tunnel Junctions

Using a tunnel junction (n^+p^+), an electrically advantageous n-n^+p^+-p-i-n configuration can be built that also may beneficially influence other structural elements (e.g. in the form of a Buried Tunnel Junction (BTJ)).

- Tunable VCSELs with micromechanically movable mirrors (MEMS): (either optically or electrically pumped).
- Wafer-bonded or wafer-fused VCSEL: Combination of semiconductor materials that can be fabricated using different types of substrate wafers.
- Monolithically optically pumped VCSELs: Two VCSELs on top of each other. One of them optically pumps the other one.
- VCSEL with longitudinally integrated monitor diode: A photodiode is integrated under the back mirror of the VCSEL. VCSEL with transversally integrated monitor diode: With suitable etching of the VCSEL's wafer, a resonant photodiode can be manufactured that may measure the light intensity of a neighboring VCSEL.

- VCSELs with external cavities (VECSELs): VECSELs are optically pumped with conventional laser diodes. This arrangement allows a larger area of the device to be pumped and therefore more power can be extracted - as much as 30W. The external cavity also allows intracavity techniques such as frequency doubling, single frequency operation and femtosecond pulse modelocking.
- Vertical-cavity semiconductor optical amplifiers: VCSOAs are optimized as amplifiers as opposed to oscillators. VCSOAs must be operated below threshold and thus require reduced mirror reflectivities for decreased feedback. In order to maximize the signal gain, these devices contain a large number of quantum wells (optically pumped devices have been demonstrated with 21–28 wells) and as a result exhibit single-pass gain values which are significantly larger than that of a typical VCSEL (roughly 5%). These structures operate as narrow linewidth (tens of GHz) amplifiers and may be implemented as amplifying filters.

Characteristics

Because VCSELs emit from the top surface of the chip, they can be tested on-wafer, before they are cleaved into individual devices. This reduces the fabrication cost of the devices. It also allows VCSELs to be built not only in one-dimensional, but also in two-dimensional arrays.

The larger output aperture of VCSELs, compared to most edge-emitting lasers, produces a lower divergence angle of the output beam, and makes possible high coupling efficiency with optical fibers.

The high reflectivity mirrors, compared to most edge-emitting lasers, reduce the threshold current of VCSELs, resulting in low power consumption. However, as yet, VCSELs have lower emission power compared to edge-emitting lasers. The low threshold current also permits high intrinsic modulation bandwidths in VCSELs.

The wavelength of VCSELs may be tuned, within the gain band of the active region, by adjusting the thickness of the reflector layers.

While early VCSELs emitted in multiple longitudinal modes or in filament modes, single-mode VCSELs are now common.

High-power VCSELs

High-power vertical-cavity surface-emitting lasers can also be fabricated, either by increasing the emitting aperture size of a single device or by combining several elements into large two-dimensional (2D) arrays. There have been relatively few reported studies on high-power VCSELs. Large-aperture single devices operating around 100 mW were first reported in 1993. Improvements in the epitaxial growth, processing, device design, and packaging led to individual large-aperture VCSELs emitting several hundreds of milliwatts by 1998. More than 2 W continuous-wave (CW) operation at -10 degrees Celsius heat-sink temperature was also reported in 1998 from a VCSEL array consisting of 1,000 elements, corresponding to a power density of 30 W/cm2. In 2001, more than 1 W CW power and 10 W pulsed power at room temperature were reported from a 19-element array. The VCSEL array chip was mounted on a diamond heat spreader, taking advantage of diamond's very high thermal conductivity. A record 3 W CW output power was reported in 2005 from large diameter single devices emitting around 980 nm.

In 2007, more than 200 W of CW output power was reported from a large (5 × 5mm) 2D VCSEL array emitting around the 976 nm wavelength, representing a substantial breakthrough in the field of high-power VCSELs. The high power level achieved was mostly due to improvements in wall-plug efficiency and packaging. In 2009, >100 W power levels were reported for VCSEL arrays emitting around 808 nm.

At that point, the VCSEL technology became useful for a variety of medical, industrial, and military applications requiring high power or high energy. Examples of such applications are:

- Medical/cosmetics: laser hair removal, laser wrinkle removal.
- Infrared illuminators for military/surveillance.
- Pumping of solid-state lasers and fiber lasers.
- High-power/high-energy second harmonic generation (blue/green light).
- Laser machining: laser cutting, laser drilling, laser ablation, laser engraving.

Applications

- Optical fiber data transmission.
- Analog broadband signal transmission.
- Absorption spectroscopy (TDLAS).
- Laser printers.
- Computer mouse.
- Biological tissue analysis.
- Chip scale atomic clock.
- Lidar for cellphone cameras.
- Structured light, e.g. the "dot projector".
- Lidar for automobile collision avoid.

Quantum Cascade Laser

Quantum cascade lasers (QCLs) are semiconductor lasers that emit in the mid- to far-infrared portion of the electromagnetic spectrum and were first demonstrated by Jerome Faist, Federico Capasso, Deborah Sivco, Carlo Sirtori, Albert Hutchinson, and Alfred Cho at Bell Laboratories in 1994.

Unlike typical interband semiconductor lasers that emit electromagnetic radiation through the recombination of electron–hole pairs across the material band gap, QCLs are unipolar and laser emission is achieved through the use of intersubband transitions in a repeated stack of semiconductor multiple quantum well heterostructures, an idea first proposed in the paper "Possibility of amplification of electromagnetic waves in a semiconductor with a superlattice" by R.F. Kazarinov and R.A. Suris in 1971.

Intersubband vs. Interband Transitions

Within a bulk semiconductor crystal, electrons may occupy states in one of two continuous energy bands - the valence band, which is heavily populated with low energy electrons and the conduction band, which is sparsely populated with high energy electrons. The two energy bands are separated by an energy band gap in which there are no permitted states available for electrons to occupy. Conventional semiconductor laser diodes generate light by a single photon being emitted when a high energy electron in the conduction band recombines with a hole in the valence band. The energy of the photon and hence the emission wavelength of laser diodes is therefore determined by the band gap of the material system used.

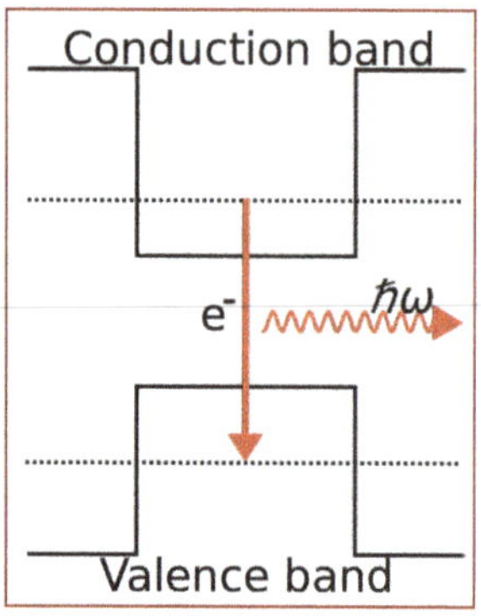

Interband transitions in conventional semiconductor lasers emit a single photon.

A QCL however does not use bulk semiconductor materials in its optically active region. Instead it consists of a periodic series of thin layers of varying material composition forming a superlattice. The superlattice introduces a varying electric potential across the length of the device, meaning that there is a varying probability of electrons occupying different positions over the length of the device. This is referred to as one-dimensional multiple quantum well confinement and leads to the splitting of the band of permitted energies into a number of discrete electronic subbands. By suitable design of the layer thicknesses it is possible to engineer a population inversion between two subbands in the system which is required in order to achieve laser emission. Because the position of the energy levels in the system is primarily determined by the layer thicknesses and not the material, it is possible to tune the emission wavelength of QCLs over a wide range in the same material system.

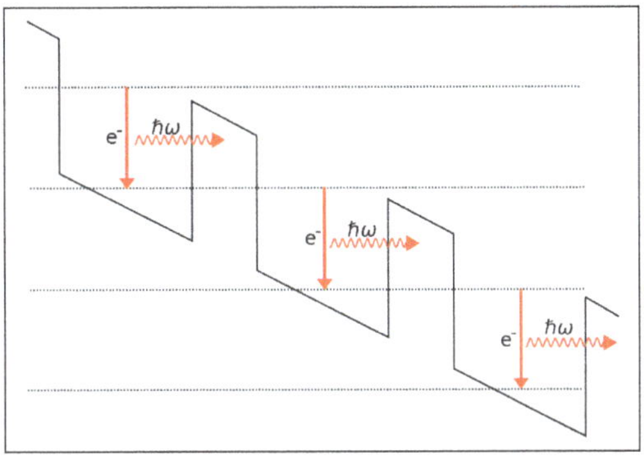

In quantum cascade structures, electrons undergo intersubband transitions and photons are emitted. The electrons tunnel to the next period of the structure and the process repeats.

Additionally, in semiconductor laser diodes, electrons and holes are annihilated after recombining across the band gap and can play no further part in photon generation. However, in a unipolar QCL, once an electron has undergone an intersubband transition and emitted a photon in one period of the superlattice, it can tunnel into the next period of the structure where another photon can be emitted. This process of a single electron causing the emission of multiple photons as it traverses through the QCL structure gives rise to the name *cascade* and makes a quantum efficiency of greater than unity possible which leads to higher output powers than semiconductor laser diodes.

Operating Principles

Rate Equations

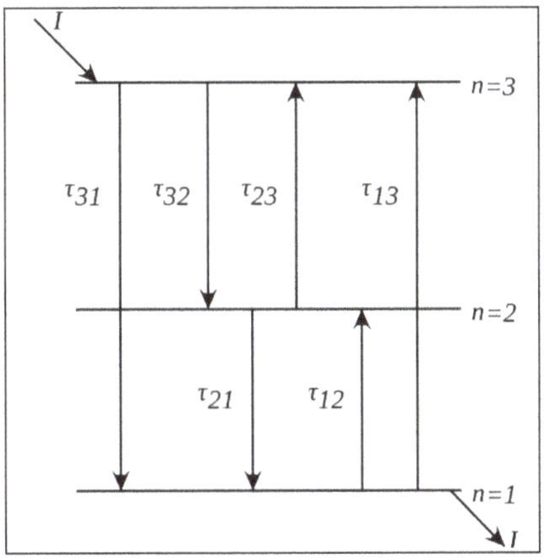

QCLs are typically based upon a three-level system. Assuming the formation of the wavefunctions is a fast process compared to the scattering between states, the time independent solutions to the Schrödinger equation may be applied and the system can be modelled using rate equations. Each subband contains a number of electrons. Each subband contains a number of electrons n_i (where i is the subband index) which scatter between levels with a lifetime T_{if} reciprocal of the average intersubband scattering rate W_{if} where i and f are the initial and final subband indices. Assuming that no other subbands are populated, the rate equations for the three level lasers are given by:

$$\frac{dn_3}{dt} = I_{in} + \frac{n_1}{\tau_{13}} + \frac{n_2}{\tau_{23}} - \frac{n_3}{\tau_{31}} - \frac{n_3}{\tau_{32}}$$

$$\frac{dn_2}{dt} = \frac{n_3}{\tau_{32}} + \frac{n_1}{\tau_{12}} - \frac{n_2}{\tau_{21}} - \frac{n_2}{\tau_{23}}$$

$$\frac{dn_1}{dt} = \frac{n_2}{\tau_{21}} + \frac{n_3}{\tau_{31}} - \frac{n_1}{\tau_{13}} - \frac{n_1}{\tau_{12}} - I_{out}$$

In the steady state, the time derivatives are equal to zero and $I_{in} = I_{out} = I$. The general rate equation for electrons in subband i of an N level system is therefore:

$$\frac{dn_i}{dt} = \sum_{j=1}^{N} \frac{n_j}{\tau_{ji}} - n_i \sum_{j=1}^{N} \frac{1}{\tau_{ij}} + I(\delta_{iN} - \delta_{i1}),$$

Under the assumption that absorption processes can be ignored, (i.e. $\frac{n_1}{\tau_{12}} = \frac{n_2}{\tau_{23}} = 0$ valid at low temperatures) the middle rate equation gives:

$$\frac{n_3}{\tau_{32}} = \frac{n_2}{\tau_{21}}$$

Therefore, if $\tau_{32} > \tau_{21}$ then $n_3 > n_2$ and a population inversion will exist. The population ratio is defined as:

$$\frac{n_3}{n_2} = \frac{\tau_{32}}{\tau_{21}} = \frac{W_{21}}{W_{32}}$$

If all N steady-state rate equations are summed, the right hand side becomes zero, meaning that the system is underdetermined, and it is possible only to find the relative population of each subband. If the total sheet density of carriers N_{2D} in the system is also known, then the *absolute* population of carriers in each subband may be determined using:

$$\sum_{i=1}^{N} n_i = N_{2D}$$

As an approximation, it can be assumed that all the carriers in the system are supplied by doping. If the dopant species has a negligible ionisation energy then N_{2D} is approximately equal to the doping density.

Active Region Designs

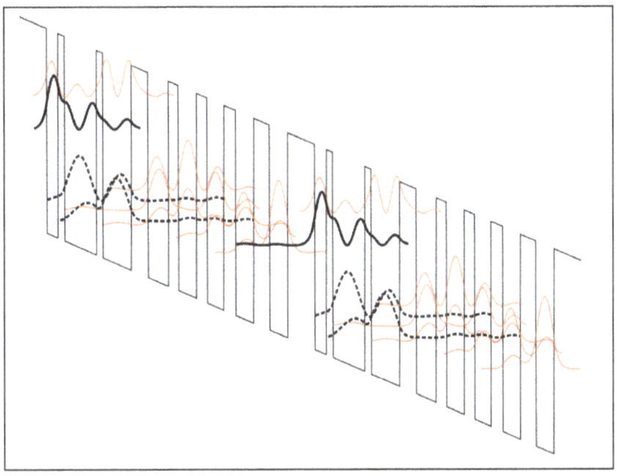

Electron wave functions are repeated in each period of a three quantum well QCL active region. The upper laser level is shown in bold.

The scattering rates are tailored by suitable design of the layer thicknesses in the superlattice which determine the electron wave functions of the subbands. The scattering rate between two subbands is heavily dependent upon the overlap of the wave functions and energy spacing between the subbands. The figure shows the wave functions in a three quantum well (3QW) QCL active region and injector.

In order to decrease W_{32}, the overlap of the upper and lower laser levels is reduced. This is often achieved through designing the layer thicknesses such that the upper laser level is mostly localised in the left-hand well of the 3QW active region, while the lower laser level wave function is made to mostly reside in the central and right-hand wells. This is known as a *diagonal* transition. A *vertical* transition is one in which the upper laser level is localised in mainly the central and right-hand wells. This increases the overlap and hence W_{32} which reduces the population inversion, but it increases the strength of the radiative transition and therefore the gain.

In order to increase W_{21}, the lower laser level and the ground level wave functions are designed such that they have a good overlap and to increase W_{21} further, the energy spacing between the subbands is designed such that it is equal to the longitudinal optical (LO) phonon energy (~36 meV in GaAs) so that resonant LO phonon-electron scattering can quickly depopulate the lower laser level.

Material Systems

The first QCL was fabricated in the GaInAs/AlInAs material system lattice-matched to an InP substrate. This particular material system has a conduction band offset (quantum well depth) of 520 meV. These InP-based devices have reached very high levels of performance across the mid-infrared spectral range, achieving high power, above room-temperature, continuous wave emission.

In 1998 GaAs/AlGaAs QCLs were demonstrated by Sirtori *et al.* proving that the QC concept is not restricted to one material system. This material system has a varying quantum well depth depending on the aluminium fraction in the barriers. Although GaAs-based QCLs have not matched the performance levels of InP-based QCLs in the mid-infrared, they have proven to be very successful in the terahertz region of the spectrum.

The short wavelength limit of QCLs is determined by the depth of the quantum well and recently QCLs have been developed in material systems with very deep quantum wells in order to achieve short wavelength emission. The InGaAs/AlAsSb material system has quantum wells 1.6 eV deep and has been used to fabricate QCLs emitting at 3.05 μm. InAs/AlSb QCLs have quantum wells 2.1 eV deep and electroluminescence at wavelengths as short as 2.5 μm has been observed.

QCLs may also allow laser operation in materials traditionally considered to have poor optical properties. Indirect bandgap materials such as silicon have minimum electron and hole energies at different momentum values. For interband optical transitions, carriers change momentum through a slow, intermediate scattering process, dramatically reducing the optical emission intensity. Intersubband optical transitions however, are independent of the relative momentum of conduction band and valence band minima and theoretical proposals for Si/SiGe quantum cascade emitters have been made.

Emission Wavelengths

QCLs currently cover the wavelength range from 2.63 μm to 250 μm (and extends to 355 μm with the application of a magnetic field).

Optical Waveguides

The first step in processing quantum cascade gain material to make a useful light-emitting device is to confine the gain medium in an optical waveguide. This makes it possible to direct the emitted light into a collimated beam, and allows a laser resonator to be built such that light can be coupled back into the gain medium.

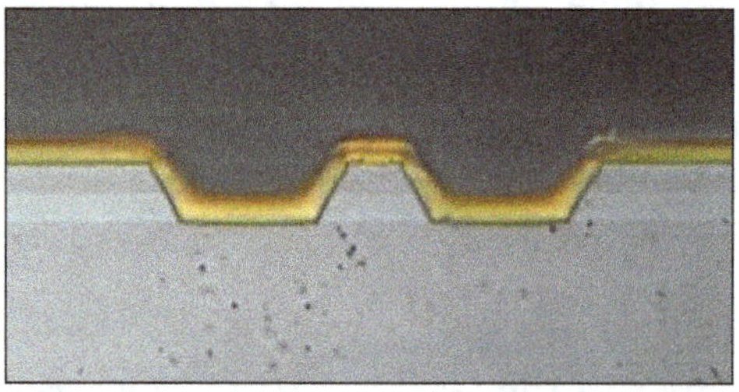

End view of QC facet with ridge waveguide. Darker gray: InP, lighter gray: QC layers, black: dielectric, gold: Au coating. Ridge ~ 10 um wide.

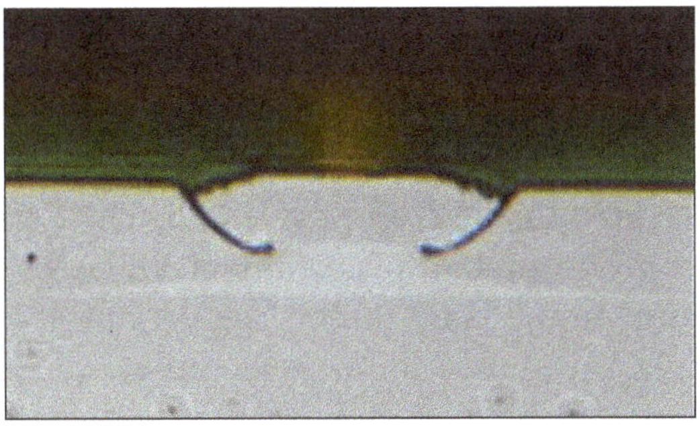

End view of QC facet with buried heterostructure waveguide. Darker gray: InP, lighter gray: QC layers, black: dielectric. Heterostructure ~ 10 um wide.

Two types of optical waveguides are in common use. A ridge waveguide is created by etching parallel trenches in the quantum cascade gain material to create an isolated stripe of QC material, typically ~10 um wide, and several mm long. A dielectric material is typically deposited in the trenches to guide injected current into the ridge, then the entire ridge is typically coated with gold to provide electrical contact and to help remove heat from the ridge when it is producing light. Light is emitted from the cleaved ends of the waveguide, with an active area that is typically only a few micrometers in dimension.

The second waveguide type is a buried heterostructure. Here, the QC material is also etched to produce an isolated ridge. Now, however, new semiconductor material is grown over the ridge. The change in index of refraction between the QC material and the overgrown material is sufficient to create a waveguide. Dielectric material is also deposited on the overgrown material around QC ridge to guide the injected current into the QC gain medium. Buried heterostructure waveguides are efficient at removing heat from the QC active area when light is being produced.

Laser Types

Although the quantum cascade gain medium can be used to produce incoherent light in a superluminescent configuration, it is most commonly used in combination with an optical cavity to form a laser.

Fabry–Perot Lasers

This is the simplest of the quantum cascade lasers. An optical waveguide is first fabricated out of the quantum cascade material to form the gain medium. The ends of the crystalline semiconductor device are then cleaved to form two parallel mirrors on either end of the waveguide, thus forming a Fabry–Perot resonator. The residual reflectivity on the cleaved facets from the semiconductor-to-air interface is sufficient to create a resonator. Fabry–Perot quantum cascade lasers are capable of producing high powers, but are typically multi-mode at higher operating currents. The wavelength can be changed chiefly by changing the temperature of the QC device.

Distributed Feedback Lasers

A distributed feedback (DFB) quantum cascade laser is similar to a Fabry–Pérot laser, except for a distributed Bragg reflector (DBR) built on top of the waveguide to prevent it from emitting at other than the desired wavelength. This forces single mode operation of the laser, even at higher operating currents. DFB lasers can be tuned chiefly by changing the temperature, although an interesting variant on tuning can be obtained by pulsing a DFB laser. In this mode, the wavelength of the laser is rapidly "chirped" during the course of the pulse, allowing rapid scanning of a spectral region.

External Cavity Lasers

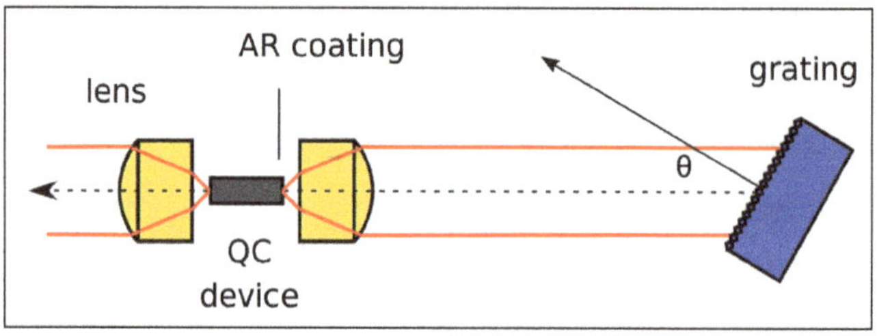

Schematic of QC device in external cavity with frequency selective optical feedback provided by diffraction grating in Littrow configuration.

In an external cavity (EC) quantum cascade laser, the quantum cascade device serves as the laser gain medium. One, or both, of the waveguide facets has an anti-reflection coating that defeats the optical cavity action of the cleaved facets. Mirrors are then arranged in a configuration external to the QC device to create the optical cavity.

If a frequency-selective element is included in the external cavity, it is possible to reduce the laser emission to a single wavelength, and even tune the radiation. For example, diffraction gratings have been used to create a tunable laser that can tune over 15% of its center wavelength.

Extended Tuning Devices

There exists several methods to extend the tuning range of quantum cascade lasers using only monolithically integrated elements. Integrated heaters can extend the tuning range at fixed operation temperature to 0.7% of the central wavelength and superstructure gratings operating through the Vernier effect can extend it to 4% of the central wavelength, compared to <0.1% for a standard DFB device.

Growth

The alternating layers of the two different semiconductors which form the quantum heterostructure may be grown on to a substrate using a variety of methods such as molecular beam epitaxy (MBE) or metalorganic vapour phase epitaxy (MOVPE), also known as metalorganic chemical vapor deposition (MOCVD).

Applications

Fabry-Perot (FP) quantum cascade lasers were first commercialized in 1998, Distributed feedback (DFB) devices were first commercialized in 2004, and broadly-tunable external cavity quantum cascade lasers first commercialized in 2006. The high optical power output, tuning range and room temperature operation make QCLs useful for spectroscopic applications such as remote sensing of environmental gases and pollutants in the atmosphere and security. They may eventually be used for vehicular cruise control in conditions of poor visibility, collision avoidance radar, industrial process control, and medical diagnostics such as breath analyzers. QCLs are also used to study plasma chemistry.

When used in multiple-laser systems, intrapulse QCL spectroscopy offers broadband spectral coverage that can potentially be used to identify and quantify complex heavy molecules such as those in toxic chemicals, explosives, and drugs.

Interband Cascade Laser

Interband cascade lasers (ICLs) are a type of laser diode that can produce coherent radiation over a large part of the mid-infrared region of the electromagnetic spectrum. They are fabricated from epitaxially-grown semiconductor heterostructures composed of layers of indium arsenide (InAs), gallium antimonide (GaSb), aluminum antimonide (AlSb), and related alloys. These lasers are similar to quantum cascade lasers (QCLs) in several ways. Like QCLs, ICLs employ the concept of bandstructure engineering to achieve an optimized laser design and reuse injected electrons to emit multiple photons. However, in ICLs, photons are generated with interband transitions, rather than the intersubband transitions used in QCLs. Consequently, the rate at which the carriers injected into the upper laser subband thermally relax to the lower subband is determined by interband Auger, radiative, and Shockley-Read carrier recombination. These processes typically occur on a much slower time scale than the longitudinal optical phonon interactions that mediates the intersubband relaxation of injected electrons in mid-IR QCLs. The use of interband transitions allows laser action in ICLs to be achieved at lower electrical input powers than is possible with QCLs.

The basic concept of an ICL was proposed by Rui Q. Yang in 1994. The key insight he had was that the incorporation of a type-II heterostructure similar to those used in interband resonant tunneling diodes would facilitate the possibility of cascade lasers that use interband transitions for photon generation. Further improvement to the design and development of the technology was carried out by Yang and his collaborators at several institutions, as well as by groups at the Naval Research Laboratory and other institutions. ICLs lasing in continuous wave (cw) mode at room temperature were first demonstrated in 2008. This laser had an emission wavelength of 3.75 μm.

Subsequently, cw operation of ICLs at room temperature has been demonstrated with emission wavelengths ranging from 2.9 μm to 5.7 μm. ICLs at cooler temperatures have been demonstrated with emission wavelengths between 2.7 μm to 11.2 μm. ICLs operating in cw mode at ambient temperature are able to achieve lasing at much lower input powers than competing mid-IR semiconductor laser technologies.

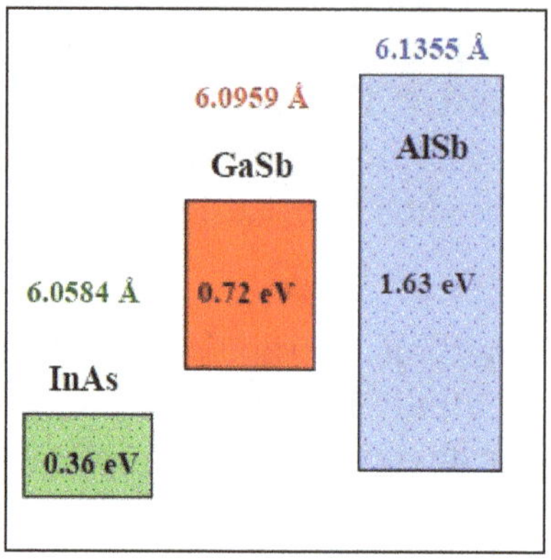

Band alignment of and lattice constant of materials used in interband cascade laser.

Theory of Operation

In a standard multiple quantum well laser, the active quantum wells used to generate photons are connected in parallel. Consequently, a large current is required to replenish each active well with electrons as it emits light. In a cascade laser, the wells are connected in series, meaning that the voltage is higher but the current is lower. This tradeoff is beneficial because the input power dissipated by the device's series resistance, R_s, is equal to I^2R_s, where I is the electric current flowing through the device. Thus, the lower current in a cascade laser results in less power loss from the device's series resistance. However, devices with more stages tend to have poorer thermal performance, since more heat is generated in locations farther from the heat sink. The optimal number of stages depends on the wavelength, material used, and several other factors. The optimization of this number is guided by simulations, but ultimately determined empirically by studying the experimental laser performance.

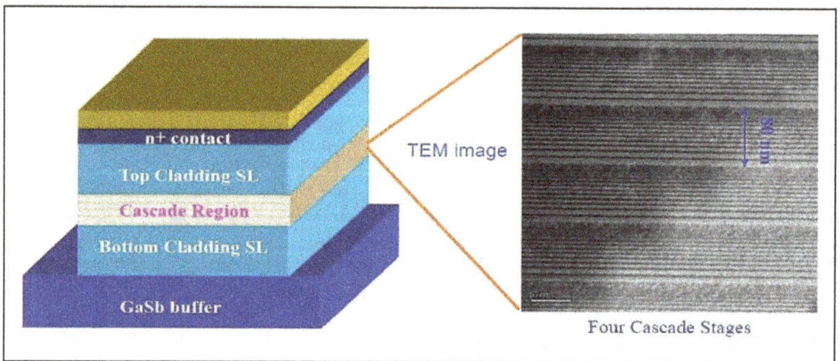

Schematic of overall epitaxial structure for laser grown on GaSb. The microscope image shows four of the thin-layer cascade stages.

ICLs are fabricated from semiconductor heterostructures grown using molecular beam epitaxy (MBE). The materials used in the structure are InAs, GaSb, AlSb, and related alloys. These three binary materials are very closely lattice matched with lattice parameters close to 6.1 Å. Thus, these materials can be incorporated together in the same heterostructure without introducing a significant amount of strain. The MBE growth is typically done on either a GaSb or InAs substrate.

The entire epitaxial structure consists of several cascade stages that are sandwiched between two separate confinement layers (SCLs), with other materials enclosing the SCLs to provide optical cladding. In addition to producing light, the layered epitaxial structure must also act as a waveguide so that the cascade stages amplify guided optical modes.

Cascade Stage Design

In each cascade stage, the thin InAs layers act as confined quantum well (QW) layers for electrons and barriers for holes. The GaSb (or GaInSb) layers conversely act as QWs for holes and barriers for electrons, while the AlSb layers serve as barriers for both electrons and holes. The key feature that enables the realization of cascading within an interband diode is the so-called "type-II", or broken-gap, band alignment between InAs and GaSb. Whereas in the more usual class of type-I QWs both the electrons and holes are confined within the same material layer, the InAs-GaSb system is type-II because the conduction band minimum of InAs lies at a lower energy than the valence band maximum of GaSb. This less common arrangement makes it easy to re-inject electrons from the valence band of one stage of the ICL into the conduction band of the next stage via simple elastic scattering.

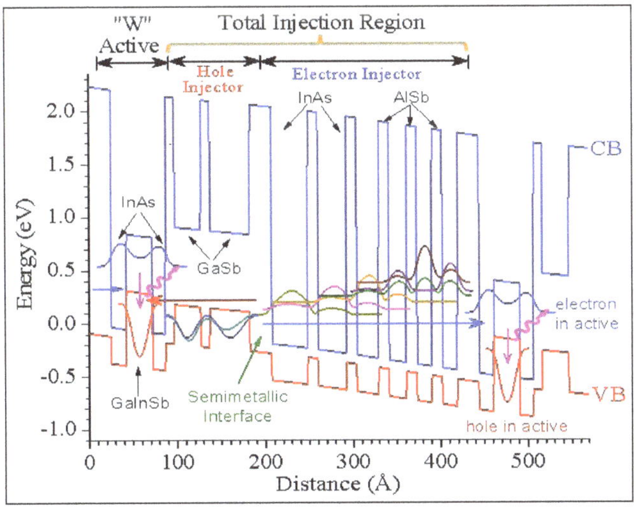

Band diagram of a single stage in a typical interband cascade laser. The cascade stage is divided into an active region, electron injector, and hole injector. The groups of quantum wells that constitute each region are indicated. The subband extrema energies and corresponding squared wavefunctions are plotted for those subbands most relevant to the device transport and laser action.

Each cascade stage effectively acts as an individual photon generator. A single stage is composed of an electron injector, a hole injector, and an active gain region consisting of one hole QW and one or two electron QWs. When the device is biased, excess electrons and holes are generated and flow into the active region, where they recombine and emit light. In order to minimize optical losses at the semimetallic interface forming the boundary between the electron and hole injectors, a layer of AlSb is placed between the InAs and GaSb layers to prevent interband reabsorption of the generated photons.

A typical active region employs the so-called "W" quantum well configuration. In this design, the GaInSb hole QW is sandwiched between two InAs electron QWs, which are in turn surrounded by two AlSb barrier layers. This arrangement maximizes the optical gain by increasing the spatial overlap between the electron and hole wavefunctions that are nominally separated in different layers. The lasing wavelength, as determined by the bandgap created between the ground state electron and hole energy levels, can be varied simply by changing the InAs electron QW thickness (whereas it is much less sensitive to the hole QW thickness).

The two injector regions are each designed to efficiently transfer its namesake carriers (electrons or holes) from the semimetallic interface to the active region. They must also double as rectifying barriers for the opposite type of carrier in order to prevent inter-stage leakage currents. The total injector (electron injector plus hole injector) should also be sufficiently thick overall to prevent the electric fields forming under bias from becoming great enough to induce dielectric breakdown of the material. The electron injector is usually made longer because of the relatively fast inter-well scattering rate of electrons compared to that of holes. This ensures a smaller series resistance contribution from the total injector transport. The hole injector is composed of GaSb/AlSb quantum wells. It is made just thick enough (typically with just one or two wells) to ensure effective suppression of electron tunneling from the active region to the electron injector of the next stage. The electron injector typically consists of a longer series of InAs/AlSb quantum wells. To maximize the InAs/AlSb superlattice miniband width, the InAs layer thicknesses are varied across the injector so that their ground state energies nearly align when the device is biased. The quantum well energy gaps in the injector must be large enough to preclude reabsorption of the photons generated by the active quantum wells.

An additional feature that differentiates the ICL from all other laser diodes is its provision for electrically-pumped operation without a p-n junction. This is possible because the injectors function as rectifying barriers that keep the current flowing in a single direction. Nevertheless, it is highly advantageous to dope certain layers in each cascade stage as a means of controlling the active electron and hole densities, via a design technique called "carrier rebalancing." While the most favorable combination of electron and hole populations depends on the relative strengths of various free carrier absorption and Auger recombination processes, the studies done thus far indicate that the ICL performance is optimal when at threshold the two concentrations are roughly equal.

Since the hole population tends to substantially exceed the electron population in undoped or moderately-doped ICLs, carrier rebalancing is achieved by heavily n-doping the electron injector (typically, with Si) so as to add electrons to the active QWs.

Optical Waveguide

The gain within a given waveguide required to reach the lasing threshold is given by the equation:

$$g_{th} = \frac{\alpha_{wg} + \alpha_{mirr}}{\tilde{A}}$$

where α_{wg} is the waveguide loss, α_{mirr} is the mirror loss, and Γ is the optical confinement factor. The mirror loss is due to photons escaping through the mirrors of the optical resonator. Waveguide losses can be due to absorption in the active, separate confinement, optical cladding materials, and metal contacts (if the claddings are not thick enough), or result from scattering at the ridge sidewalls. The confinement factor is that percentage of the optical energy concentrated in the cascade stages. As with other semiconductor lasers, ICLs have a tradeoff between optical loss in the waveguide and Γ. The overall goal of waveguide design is to find the proper structure that minimizes the threshold gain.

The choice of waveguide material depends on the substrate used. For ICLs grown on GaSb, the separate confinement layers are typically low-doped GaSb while the optical cladding layers are InAs/AlSb superlattices lattice-matched to the GaSb substrate. The bottom cladding must be fairly thick to prevent leakage of the guided mode into the substrate, since the refractive index of GaSb is larger than the effective index of the lasing mode.

An alternative waveguide configuration that is suitable for growth on InAs substrates uses highly *n*-doped InAs for the optical cladding. The high electron density in this layer lowers the refractive index in accordance with the Drude model. In this approach, the epitaxial structure is grown on an *n*-type InAs substrate and it also utilizes InAs for the separate confinement layers. For longer-wavelength operation, advantages include the much higher thermal conductivity of bulk InAs as compared to a short-period InAs/AlSb superlattice, as well as a much thinner cladding layer due its larger index contrast with the active region. This shortens the MBE growth time, and also further improves the thermal dissipation. However, the waveguide must be designed carefully to avoid excessive free carrier absorption loss in the heavily-doped layers.

Current Status of ICL Performance

ICLs emitting at 3.7 um have operated in cw mode up to a maximum temperature of 118 °C. A maximum cw output power of nearly 0.5 W has been demonstrated at room temperature, with 200-300 mW in a nearly-diffraction-limited beam. A maximum room-temperature cw wall-plug efficiency of nearly 15% has also been achieved. While QCLs typically require input electrical powers of nearly 1 W and higher to operate at room temperature, ICLs are able to lase for input powers as low as 29 mW owing to the much longer interband carrier lifetime. Room-temperature cw operation with low dissipated powers can be achieved for wavelengths between approximately 3.0 um and 5.6 um.

The figure on the right shows the performance characteristics of narrow ridge-waveguide interband cascade lasers at room temperature operating in cw mode. Specifically, the figure shows plots of the amount of power emitted by lasers with different ridge widths for a given injection current. Each of these lasers had five cascade stages and cavity lengths of 4 mm. These lasers were mounted so that the top of the epitaxial structure (rather than the substrate) was in contact with the copper heat sink (typically referred to as an epitaxial side down configuration) in order to achieve optimal heat dissipation. In addition, they were fabricated with corrugated sidewalls. The sidewall corrugation lowers optical losses by ensuring fewer photons are generated in the higher-order optical modes that are more susceptible to optical scattering losses.

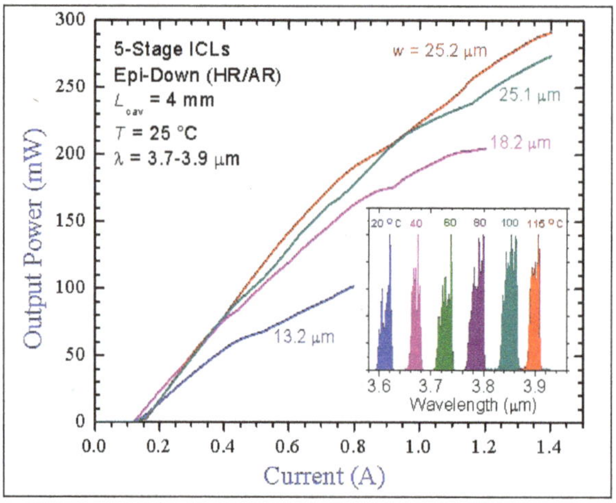

Light-current characteristics in continuous-wave mode at room temperature for narrow ridge-waveguide interband cascade lasers with several different ridge widths (w) as indicated in the figure.

At the maximum output power, the beam quality is within ≈2 times the diffraction limit for all the ridges. The cw lasing wavelength of these ICLs span from 3.6 to 3.9 µm in temperature range from 20 to 115 °C (as shown in inset).

Applications

Mid-infrared lasers are important tools for spectroscopic sensing applications. Many molecules such as those in pollution and greenhouse gases have strong rotational and vibrational resonances in the mid-infrared region of the spectrum. For most sensing applications, the laser wavelength must also be within one of the atmospheric window to avoid signal attenuation.

An important requirement for this type of application is that single-mode emission is obtained. With ICLs, this can be done by making distributed feedback lasers. A distributed-feedback ICL, designed for the excitation of methane gas, was developed at NASA Jet Propulsion Laboratory and included as an instrument on the tunable laser spectrometer on the Curiosity rover that was sent to explore the environment of Mars. A more recent distributed feedback ICL emitted up to 27 mW in a single spectral mode at 3.79 μm when operated at 40 °C, and 1 mW for operation at 80 °C.

Quantum Well Laser

A quantum well laser is a laser diode in which the active region of the device is so narrow that quantum confinement occurs. Laser diodes are formed in compound semiconductor materials that (quite unlike silicon) are able to emit light efficiently. The wavelength of the light emitted by a quantum well laser is determined by the width of the active region rather than just the bandgap of the material from which it is constructed. This means that much longer wavelengths can be obtained from quantum well lasers than from conventional laser diodes using a particular semiconductor material. The efficiency of a quantum well laser is also greater than a conventional laser diode due to the stepwise form of its density of states function.

In 1972, Charles H. Henry, a physicist and newly appointed Head of the Semiconductor Electronics Research Department at Bell Laboratories, had a keen interest in the subject of integrated optics, the fabrication of optical circuits in which the light travels in waveguides.

Later that year while pondering the physics of waveguides, Henry had a profound insight. He realized that a double heterostructure is not only a waveguide for light waves, but simultaneously for electron waves. Henry was drawing upon the principles of quantum mechanics, according to which electrons behave both as particles and as waves. He perceived a complete analogy between the confinement of light by a waveguide and the confinement of electrons by the potential well that is formed from the difference in bandgaps in a double heterostructure.

C.H. Henry realized that, just as there are discrete modes in which light travels within a waveguide, there should be discrete electron wavefunction modes in the potential well, each having a unique energy level. His estimate showed that if the active layer of the heterostructure is as thin as several tens of nanometers, the electron energy levels would be split apart by tens of milli-electron volts. This amount of energy level splitting is observable. The structure Henry analyzed is today called a "quantum well."

Henry proceeded to calculate how this "quantization" (i.e., the existence of discrete electron wavefunctions and discrete electron energy levels) would alter the optical absorption properties (the absorption "edge") of these semiconductors. He realized that, instead of the optical absorption increasing smoothly as it does in ordinary semiconductors, the absorption of a thin heterostructure (when plotted versus photon energy) would appear as a series of steps.

In addition to Henry's contributions, the quantum well (which is a type of double-heterostructure laser) was actually first proposed in 1963 by Herbert Kroemer in Proceedings of the IEEE and simultaneously in the U.S.S.R by Zh. I. Alferov and R.F. Kazarinov. Alferov and Kroemer shared a Nobel Prize in 2000 for their work in semiconductor heterostructures.

Experimental Verification of Quantum Wells

In early 1973, Henry proposed to R. Dingle, a physicist in his department, that he look for these predicted steps. The very thin heterostructures were made by W. Wiegmann using molecular beam epitaxy.

Invention of the Quantum Well Laser

After this experiment showed the reality of the predicted quantum well energy levels, Henry tried to think of an application. He realized that the quantum well structure would alter the density of states of the semiconductor, and result in an improved semiconductor laser requiring fewer electrons and electron holes to reach laser threshold. Also, he realized that the laser wavelength could be changed merely by changing the thickness of the thin quantum well layers, whereas in the conventional laser a change in wavelength requires a change in layer composition. Such a laser, he reasoned, would have superior performance characteristics compared to the standard double heterostructure lasers being made at that time.

Dingle and Henry received a patent on this new type of semiconductor laser comprising a pair of wide bandgap layers having an active region sandwiched between them, in which "the active layers are thin enough (e.g., about 1 to 50 nanometres) to separate the quantum levels of electrons confined therein. These lasers exhibit wavelength tunability by changing the thickness of the active layers. Also described is the possibility of threshold reductions resulting from modification of the density of electron states."

Quantum well lasers require fewer electrons and holes to reach threshold than conventional double heterostructure lasers. A well-designed quantum well laser can have an exceedingly low threshold current.

Moreover, since quantum efficiency (photons-out per electrons-in) is largely limited by optical absorption by the electrons and holes, very high quantum efficiencies can be achieved with the quantum well laser.

To compensate for the reduction in active layer thickness, a small number of identical quantum wells are often used. This is called a multi-quantum well laser.

Early Demonstrations

While the term "quantum well laser" was coined in the late 1970s by Nick Holonyak and his students at the University of Illinois at Urbana Champaign, the first observation of quantum well laser operation was made in 1975 at Bell Laboratories. The first electrically pumped "injection" quantum well laser was observed by P. Daniel Dapkus and Russell D. Dupuis of Rockwell International, in collaboration with the University of Illinois at Urbana Champaign (Holonyak) group in 1977. Dapkus and Dupuis had, by then, pioneered the metalorganic vapour phase epitaxy MOVPE (also known as OMCVD, OMVPE, and MOCVD) technique for fabricating semiconductor layers. The MOVPE technique, at the time, provided superior radiative efficiency as compared to the molecular beam epitaxy (MBE) used by Bell Labs. Later, however, Won T. Tsang at Bell Laboratories succeeded in using MBE techniques in the late 1970s and early 1980s to demonstrate dramatic improvements in performance of quantum well lasers. Tsang showed that, when quantum wells are optimized, they have exceedingly low threshold current and very high efficiency in converting current-in to light-out, making them ideal for widespread use.

The original 1975 demonstration of optically pumped quantum well lasers had threshold power density of 35 kW/cm^2. Ultimately, it was found that the lowest practical threshold current density in any quantum well laser is 40 Amperes/cm^2, a reduction of approximately 1,000x.

Extensive work has been performed on quantum well lasers based on gallium arsenide and indium phosphide wafers. Today, however, lasers utilizing quantum wells and the discrete electron modes researched by C.H. Henry during the early 1970s, fabricated by both MOVPE and MBE techniques, are produced at a variety of wavelengths from the ultraviolet to the THz regime. The shortest wavelength lasers rely on gallium nitride-based materials. The longest wavelength lasers rely on the quantum cascade laser design.

Creation of the Internet

Quantum well lasers are important because they are the basic active element (laser light source) of the Internet fiber optic communication. Early work on these lasers focused on GaAs gallium arsenide based wells bounded by Al-GaAs walls, but wavelengths transmitted by optical fibers are best achieved with indium phosphide walls with indium gallium arsenide phosphide based wells. The central practical issue of light sources buried in cables is their lifetimes to burn-out. The average burn-out time of early quantum well lasers was less than one second, so that many early scientific successes were achieved using rare lasers with burn-out times of days or weeks. Commercial success was achieved by Lucent (a spin-off from Bell Laboratories) in the early 1990's with quality control of quantum well laser production by MOVPE Metalorganic vapour phase epitaxy, as done using high-resolution X rays by Joanna (Joka) Maria Vandenberg. Her quality control produced Internet lasers with median burn-out times longer than 25 years.

Vertical-external-cavity Surface-emitting-laser

A vertical-external-cavity surface-emitting-laser (VECSEL) is a small semiconductor laser similar to a vertical-cavity surface-emitting laser (VCSEL). VECSELs are used primarily as near infrared devices in laser cooling and spectroscopy, but have also been explored for applications such as telecommunications.

Comparisons with VCSELs

Unlike a VCSEL, in which two high-reflecting mirrors are incorporated into the laser structure to form the optical cavity, in a VECSEL one of the two mirrors is external to the diode structure. As a result, the cavity includes a free-space region. A typical distance from the diode to the external mirror would be 1 cm. Several workers demonstrated optically pumped VECSELs, and they continue to be developed for many applications including very high power diode laser sources for use in industrial machining (cutting, punching, etc.) because of their unusually high power and efficiency when pumped by multi-mode diode laser bars. These lasers are in the process of challenging conventional high power lasers such as solid state (e.g., Nd:YAG) and carbon dioxide lasers for machining operations.

However, electrically pumped VECSELs (another matter entirely), were the brainchild of Aram Mooradian, an engineer known for fundamental contributions to diode laser linewidth studies, who worked for many years at MIT Lincoln Laboratory in Lexington, Massachusetts. Mooradian formed a company, Novalux, Inc., which was the first to demonstrate VECSELs (which they called "NECSELs"). Applications for electrically pumped VECSELs include frequency doubling of near-IR VECSEL emitters to attain compact powerful sources of single-mode blue and green light for projection display purposes.

Semiconductor Gain

One of the most interesting features of any VECSEL is the thinness of the semiconductor gain region in the direction of propagation, less than 100 nm. In contrast, a conventional in-plane semiconductor laser entails light propagation over distances of from 250 μm upward to 2 mm or longer. The significance of the short propagation distance is that it causes the effect of "antiguiding" nonlinearities (the same phenomenon is coincidentally quantified by the linewidth enhancement factor relating to Mooradian's above-mentioned earlier work) in the diode laser gain region to be minimized. The result is a large-cross-section single-mode optical beam which is not attainable from in-plane (a.k.a. "edge-emitting") diode lasers.

In a VECSEL, the external mirror permits a significantly greater area of the diode to participate in generating light in a single mode, resulting in much higher power than otherwise attainable. Monolithic VCSELs emit powers in the low milliwatt range.

By contrast, at the 2004 Optical Society of America "Conference on Lasers and Electro-Optics," held in San Francisco, California, one company (Coherent, Inc.) announced 45 watt continuous wave single-mode emission from an optically pumped VECSEL. Numerous other companies and organizations worldwide have adopted the optically pumped architecture for its simplicity.

Distributed Bragg Reflector Laser

A *distributed Bragg reflector laser* is a laser, where the laser resonator is made with at least one distributed Bragg reflector (DBR) outside the gain medium (the active region). A DBR is a Bragg mirror, i.e., a light-reflecting device (a mirror) based on Bragg reflection at a periodic structure. In most cases, the Bragg mirror is more specifically a quarter-wave mirror, providing the maximum amount of reflection for the given number of layers.

Lasers of DBR type are usually laser diodes, but the term is also sometimes used for fiber lasers containing fiber Bragg gratings. Both laser types are described below. Most solid-state bulk lasers actually also use laser mirrorswhich are Bragg mirrors; nevertheless such lasers are not called DBR lasers.

A DBR laser is different from a distributed feedback laser, where the whole active medium is embedded in a single distributed reflector structure.

DBR Laser Diodes

A DBR laser diode contains some corrugated waveguide structure (a grating section) providing wavelength-dependent feedback to define the emission wavelength. Another section of the laser waveguide acts as the amplifying medium (active region), and the other end of the resonator may have another DBR.

DBR laser diodes are usually single-frequency lasers with diffraction-limited output, and often they are wavelength-tunable (→ *tunable lasers*). Tuning within the free spectral range of the laser resonator may be accomplished with a separate *phase section*, which can e.g. be electrically heated, or simply by varying the temperature of the gain region via the drive current. If the temperature of the whole device is varied, the wavelength response is significantly smaller than for an ordinary single-mode laser diode, since the reflection band of the grating is shifted less than the gain maximum. Electro-optic tuning can also be accomplished. Mode-hop free tuning over a larger wavelength region is possible by coordinated tuning of the Bragg grating and the gain structure.

There are more sophisticated device designs, exploiting a kind of Vernier effect with sampled gratings (*SG-DBR laser*), that offer a tuning range as wide as e.g. 40 nm, although not without mode hops.

The linewidth of a DBR diode is typically a few megahertz. Due to the relatively short laser resonator, it is larger than that e.g. of an external-cavity diode laser.

There are MOPA structures where an additional amplifier section (a semiconductor optical amplifier) is placed on the same semiconductor chip. The actual DBR laser is then the seed laser. Output powers well above 100 mW can be achieved with such devices. It is also possible to directly generate high powers (> 10 W) with a broad-area laser diode-having a surface Bragg grating within its resonator. The emission may then no longer be in a single mode, but still with a relatively small bandwidth.

Vertical cavity surface-emitting lasers (VCSELs) are actually also distributed Bragg reflector lasers, even though the term "DBR laser diodes" is normally used for edge-emitting semiconductor lasers.

Applications of DBR laser diodes include optical fiber communications, free-space optical communications, laser cooling, optical metrology and sensors, and high-resolution laser spectroscopy. DBR lasers actually compete with external-cavity diode lasers (ECDLs), which also offer wavelength-tunable single-frequency output, with potentially better performance e.g. in terms of noise, but also requiring a significantly more complex setup. Chips containing DBR laser arrays can serve as very compact sources for use in wavelength division multiplexing systems.

DBR Fiber Lasers

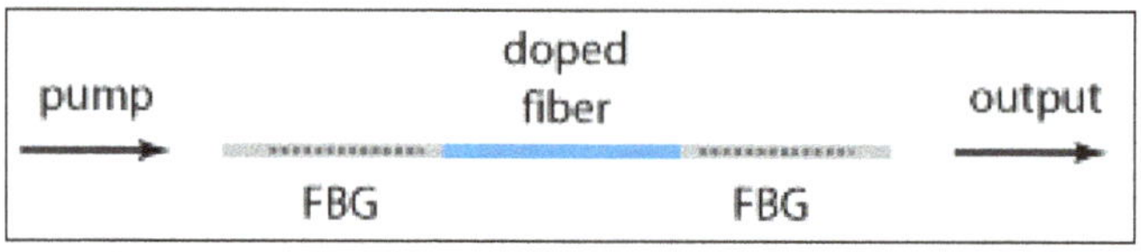

Short DBR fiber laser for narrow-linewidth emission.

A fiber laser of DBR type usually has a linear laser resonator formed by an active (rare-earth-doped) fiber between two fiber Bragg gratings. Compared with a fiber DFB laser, which consists of a single grating in a fiber with laser gain, a DBR fiber laser has a longer laser resonator and thus the potential for higher output power, higher power efficiency, and narrower linewidth. On the other hand, this can also lead to less robust single-frequency operation, or to multimode operation with a correspondingly much larger emission bandwidth. Single-frequency DBR fiber lasers offer similar output powers as DBR laser diodes: tens of milliwatts or sometimes > 100 mW.

Quantum Dot Laser

A quantum dot laser is a semiconductor laser that uses quantum dots as the active laser medium in its light emitting region. Due to the tight confinement of charge carriers in quantum dots, they exhibit an electronic structure similar to atoms. Lasers fabricated from such an active media exhibit device performance that is closer to gas lasers, and avoid some of the negative aspects of device performance associated with traditional semiconductor lasers based on bulk or quantum well active media. Improvements in modulation bandwidth, lasing threshold, relative intensity noise, linewidth enhancement factor and temperature insensitivity have all been observed. The quantum dot active region may also be engineered to operate at different wavelengths by varying dot size and composition. This allows quantum dot lasers to be fabricated to operate at wavelengths previously not possible using semiconductor laser technology.

Recently, devices based on quantum dot active media are finding commercial application in medicine (laser scalpel, optical coherence tomography), display technologies (projection, laser TV), spectroscopy and telecommunications. A 10 Gbit/s quantum dot laser that is insensitive to temperature fluctuation for use in optical data communications and optical networks has been developed using this technology. The laser is capable of high-speed operation at 1.3 μm wavelengths, at temperatures from 20 °C to 70 °C. It works in optical data transmission systems, optical LANs and metro-access systems. In comparison to the performance of conventional strained quantum-well lasers of the past, the new quantum dot laser achieves significantly higher stability of temperature.

Transistor Laser

Transistor laser is a semiconductor device that functions as a transistor with an electrical output and an optical output as opposed to the typical two electrical outputs. This optical output separates it from typical transistors and, because optical signals travel faster than electrical signals, has the potential to speed up computing immensely. Researchers who discovered the transistor laser developed a new model of Kirchhoff's current law to better model the behavior of simultaneous optical and electrical output.

Construction of Transistor

The transistor laser functions like a typical transistor, but emits infrared light through one of its outputs rather than electricity. A reflective cavity within the device focuses the emitted light into a laser beam. The transistor laser is a heterojunction bipolar transistor (using different materials between the base and emitter regions) that employs a quantum well in its base region that causes emissions of infrared light. While all transistors emit some small amount of light during operation, the use of a quantum well increases the intensity of light output by as much as 40 times.

The laser output of the device works when the quantum well in the base region captures electrons that would normally be sent out through the electrical output. These electrons then undergo a process of radiative recombination, during which electrons and positively charged "holes" recombine in the base. While this process occurs in all transistors, it has an exceedingly short lifespan of only 30 picoseconds in the transistor laser, allowing for faster operation. Photons are then released through stimulated emission. Light bounces back and forth between reflective walls inside the 2.2 micrometer wide emitter that acts as a resonant cavity. Finally, light is emitted as a laser.

The device was initially constructed out of layers of indium gallium phosphide, gallium arsenide, and indium gallium arsenide, which prevented the device from running without being cooled with liquid nitrogen. Current materials allow for operation at 25 °C and continuous wave operation (continuously emitting light) at 3 GHz. The transistor laser can produce laser output without any resonance peak in the frequency response. It also does not suffer from unwanted self-resonance that results in errors in transmitted information that would necessitate complicated external circuitry to rectify.

Potential to Speed up Computers

Even though the transistor laser is still only the subject of research, there has been significant amount of speculation as to what one could be used for, especially in computing. For instance, its optical capabilities could be used to transfer data between memory chips, graphics cards, or other internal computer elements at faster rates. Currently, optic-fiber communication requires transmitters that convert electrical signals to pulses of light, and then a converter on the other end to revert these pulses back to electrical signals. This makes optical communication within computers impractical. Optical communication within computers could soon be practical, though, because the conversion of electricity to optical signals and vice versa occurs within the transistor laser without the need for external circuitry. The device could also speed up current optical communication in other applications, such as in the communication of large amounts of data over long distances.

Changing Kirchhoff's Laws

The research team that discovered the transistor laser claimed that one of Kirchhoff's laws would have to be reconstructed to include energy conservation, as opposed to just current and charge. Because the transistor laser provides two different kinds of output, the team of researchers responsible for the transistor laser had to modify Kirchhoff's current law to apply to the balance of energy as well as the balance of charge. This marked the first time Kirchhoff's laws had been extended to apply to not just electrons, but photons, too.

References

- Deppe, Herbert; Horch, Hans-Henning (2007). "Laser applications in oral surgery and implant dentistry" (PDF). Lasers in Medical Science. 22 (4): 217–221. doi:10.1007/s10103-007-0440-3. PMID 17268764

- Semiconductor-las, microwave-radar, electronics: daenotes.com, Retrieved 20 February, 2019
- Calvez, S.; Hastie, J.E.; Guina, M.; Okhotnikov, O.G.; Dawson, M.D. (2009). "Semiconductor disk lasers for the generation of visible and ultraviolet radiation". Laser & Photonics Reviews. 3 (5): 407–434. Doi:10.1002/lpor.200810042
- Vurgaftman, I.; R. Weih; M. Kamp; J.R. Meyer; C.L. Canedy; M. Kim; W.W. Bewley; C.D. Merritt; J. Abell; S. Hoefling (2015). "Topical Review - Interband cascade lasers". Journal of Physics D: Applied Physics. 48: 123001–123017.
- Distributed-bragg-reflector-lasers: p-photonics.com, Retrieved 21 March, 2019
- Then, H. W., N. Holonyak, Jr., and M. Feng. "Microwave Circuit Model of the Three-port Transistor Laser." JOURNAL OF APPLIED PHYSICS 108 (2010): n. Pag. Web

Applications in Diverse Fields

6

- Laser Applications in Defence
- Laser Applications in Civil
- Laser Applications in Surgery
- Miscellaneous Applications of Laser
- Industrial and Commercial Applications
- Laser Beam Machining
- Laser Ablation

Lasers are used in various fields such as defense, surgery, medicine and in military operations. The industrial and commercial applications of lasers include laser cutting, laser pointers, laser drilling, laser engraving and laser beam welding. This chapter has been carefully written to provide an easy understanding of these applications of lasers.

Laser Applications in Defence

Laser Range Finder

To knock down an enemy tank, it is necessary to range it very accurately. Because of its high intensity and very low divergence even after travelling quite a few kilometres, laser is ideally suited for this purpose. The laser range finders using neodymium and carbon dioxide lasers have become a standard item for artillery and tanks. These laser range finders are light weight and have higher reliability and superior range accuracy as compared to the conventional range finders.

The laser range finder works on the principle of a radar. It makes use of the characteristic properties of the laser beam, namely, monochromaticity, high intensity, coherency, and directionality. A collimated pulse of the laser beam is directed towards a target and the reflected 1ight from the target is received by an optical system and detected. The time taken by the laser beam for the to and fro travel from the transmitter to the target is measured. When half of the time thus recorded is multiplied by the velocity of light, the product gives the range, i.e., the distance of the target.

The laser range finder is superior to microwave radar as the former provides better collimation or directivity which makes high angular resolution possible. Also, it has the advantage of greater radiant brightness and the fact that this brightness is highly directional even after travelling long distances, the size of the emitting system is greatly reduced. The high mono chromaticity permits the use of optical band pass filter in the receiver circuit to discriminate between the signal and the stray light noise.

A typical laser range finder can be functionally divided into four parts: (i) transmitter, (ii) receiver, (iii) display and readout, and (iv) sighting telescope. The transmitter uses a Q-switched Nd:YAG laser which sends out single, collimated and short pulse of laser radiation to the target. A scattering wire grid directs a small sample of light from the transmitter pulse on to the photodetector, which after amplification is fed to the counter. This sample of light starts the counter. The reflected pulse, received by the telescope, is passed through an interference filter to eliminate any extraneous radiation. It is then focused on to another photodetector. The resulting signal is then fed to the counter. A digital system converts the time interval into distance. The range, thus determined by the counter, is displayed in the readout. The lighting telescope permits the operator to read the range while looking at the target.

Special circuits have been used to eliminate Spurious signals with the help of range gating and to make the use of laser range finder Possible under all weather conditions for which the targets can be seen visually through the sighting telescope. The modern versions of the laser range finders Use either high repetition pulsed Nd:YAG laser or carbon dioxide laser with range gating system. In ranging a target about 10 km away using these systems, an accuracy within 5 m is easily obtained. The laser range finders of medium range (up to 10 km) are used in several Defence areas, including laser range finder for artillery, an armoured vehicle, or a truck.

Portable laser range finders, used in the field artillery fire control systems. These are intended for field application in conjunction with artillery fire control systems.

Airborne laser range finder, pod-mounted and servo-positioned for the Air Force. In any airborne weapon system, one of the i.e., the distance of the target. The laser range finder combines the characteristic features of a laser with gyroscope stabilisation to provide an equipment which is more accurate and has a faster response than any other means of deriving air-to-surface or air- to-air range. At the same time, it is more compact than any radar.

The laser walkie-talkie range finder, a compact small instrument, weighing less than 4 kg, useful to range objects at distances less than 5 km. This range finder uses the semiconductor diode laser In emitting short duration pulses. With this, it is possible to which transmit and receive audio/visual communications, or pinpoint targets with a hand-held laser, even from unsteady environment in a helicopter or on a ship being tossed around by the rolling seas. There are no separate tripods, unwieldly power packs, or other external accessories.It gives an immediate readout of distance and elevation right on the instrument.

Underwater Laser

Lasers can also be used as a source of underwater transmission. For this purpose, a laser giving radiation in the blue-green region is most suitable as the transmission in this region is maximum for sea water. The attenuation in underwater transmission is due to (i) absorption by materials in water, (ii) scattering by suspended particles, and (iii) variation in optical density along the light path. The blue-green lasers have assumed much importance in the systems related to naval applications.

At present, the submarines have to rely on a sonar to find the enemy crafts and to avoid the underwater objects. This has serious limitations. The whales, dolphins and other marine life give false signals. A typical sonar cannot give a well-defined picture because the sonar beam is broadened or scattered by sea water. A difference in the saltiness of water can cause the sonar beam to bend and make the target appear where it is not. Another problem of using sonar is that it gives away to the enemy the position of the ship from which it is transmitted.

Lasers can be used efficiently for ranging and detection of underwater objects. For this purpose, a frequency doubled *Nd:YAG* laser or an argon ion gas laser or a Raman shifted xenon chloride laser is used. It consists of the laser transmitter which sends high power laser pulses of about 10 ns duration to the target at the rate of 30 to 50 per second through a beam splitter and a diffuser small amount of the laser light reflected by the beam splitter is made to fall on the photodiode the ranging and display circuit to start the time interval counter. The reflected light from the target is collected by telescopic optics after stray radiation is eliminated by an interference filter. A range gating circuit helps to avoid unwanted echoes The reflected pulse from the target is intensified by the image intensifier and the output is fed to image orthicon, which gives the display of the object. In this way, both the range and the image of the target are obtained. With high power release of several megawatts power, underwater ranging is possible up to 500 m in clear water.

Lasers can also be used for communication between submarines ensuring absolute privacy and in guidance systems for torpedoes and other unmanned underwater vehicles. Recent underwater laser communication has been established via satellite, i.e., from ground-to-satellite and then to underwater station.

Laser-guided Anti-tank Missile

A missile can be guided and controlled by an infrared beam emitted from a laser, with extremely small divergence. This can be achieved in four ways:

- The laser beam is used to illuminate the target tank; the anti-tank missile (ATM) then homes on to the target,as the latter has. become a source of back-scattered radiation.
- The laser beam is used to provide guidance instructions to the missile, i.e., it provides the command link.
- The missile rides the laser beam which is kept pointing along the collision course to the target.
- The missile itself carries a laser scanner and seeker for active homing on to the target.

In the first case, the laser target designator is a pulsed *Nd:YAG* laser. The laser beam is so modulated that the receiver, a four quadrant detector in the missile, is able to calculate any divergence of the missile trajectory from the beam axis and correct the deviation by altering the fins of the missile. The guidance unit consists of both optical and electronic equipment. This enables the gunner to aim the infrared guidance beam for firing the missiles.

The system in which the missile is a beam rider designed to ride the laser pointing in the direction of the target, is more attractive. n missile can carry four detectors at the wing tips looking towards the rear of the missile. The detectors determine the central axis of the laser beam and keep the flight path of the missile along it. The wavelength of the laser should be such that is the least absorbed by the plume of the sustainer motor. Thus in a laser designator, the laser by virtue of its narrow beam illuminates a chosen target. A receiver in a bomb or a missile seeks the target illuminated from the scattered laser radiation and homes on to it. In the Vietnam war an in the recent Iraqi war, the Americans used laser guided missiles with pinpoint accuracy to destroy the enemy targets.

Laser Radar

When the laser beam is used for a radar application, it is called lidar. The details, which could not be achieved earlier with microwave radars, can now be obtained with lidar. Besides, the laser beam can be focused with lenses an mirrors easily whereas microwaves need huge antenna for focusing. As a beacon or a radar, the advantages of utilising small antenna and components are obvious. With a lidar, the dimension and the distance of the target can be obtained with higher accuracy, which is not possible wit, the conventional microwave radar. The lasers used in lidars are of carbon dioxide, Q-switched neodymium, or gallium arsenide semiconductor type.

The great advantage of the use of carbon dioxide lasers for radar application is their capacity to produce high power output with requisite The spectral purity. The coherent carbon dioxide laser tips radar functions essentiaı like a coherent microwave radar except for the fact that the carbon dioxide laser beam has a frequency of a few thousand times more than that of the X-band radar and at it a sharp beam width of a few microradians. The high frequency of the carbon dioxide laser also produces high *Doppler shift* even from slow-moving targets. The fine beam width and high *Doppler* the *shift* give the carbon dioxide laser an unparalleled imaging capability. This radar system is used for and measuring radial velocities to track low-flying aircraft and slow-moving objects. Since the laser beam is very much attenuated by rain, fog, or snow, the lidar can perform well only in good weather conditions.

Ring Laser Gyroscope

The ring laser gyroscope is an extremely useful instrument for sensing and measuring very small angles of rotation of the moving objects. It has now replaced the mechanical gyroscopes used in most of the aircraft (both civil and military) he and also in long range guided missiles. The main advantages of the ring laser gyroscope are: (i) non-existence of moving parts, (ii) high g capability, and (iii) higher reliability as compared to the mechanical gyroscope. In addition, the laser gyroscope is capable of wide dynamic range and rapid reaction time, the characteristics required for missile guidance.

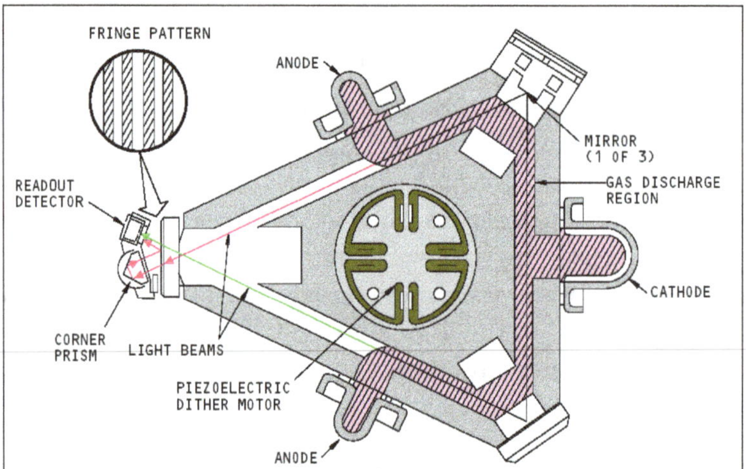

Laser Gyro

The ring laser gyroscope basically consist of a ring cavity around which two laser light beams travel in opposite directions. The operation of the ring laser gyroscopes is based on the so called Sagnac effect by which rotation of an object is sensed by an interferometric technique. A schematic diagram of the ring laser gyroscope.

In a triangular cavity of a quartz block, laser beam is split into two light beams with the help of suitable mirrors. These two light beam travel in opposite directions in the same path of the cavity, one in the clockwise and the other in the anti-clockwise direction. The two light beams then pass through a beam splitter and a beam combiner, behind which a readout detector is placed. If the cavity which is acting as an interferometer is stationary, the two light beams travel the same distance in the opposite directions without any path difference and hence no interference takes place. However, if the block is rotated clockwise about an axis through the centre and perpendicular to the plane of the interferometer, the beam travelling in the clockwise direction travels a path length slightly more than the beam travelling in the anti-clockwise direction. When these two light beams recombine at the beam combining prism, interference takes place due to the path difference; the interference fringes displaced in the field of view are proportional to the amount of rotation of the block. The laser gyroscope uses a helium-neon gas laser to generate monochromatic radiation in the two directions inside the triangular quartz block. Two photo detectors sense the direction and the rate of rotation. The output is proportional to the input. angle. The whole system is a single plane, rate integrating gyroscope and is capable of measuring rotation rates of the order of 1/10,000 degree/hour.

The main use of the ring laser gyroscope is for inertial navigation. It is being used in inertial guidance of aircraft, ships, and missiles; flight control; and gun-fire pointing. Both Honeywell and a Litton Industries, USA, the manufacturers of the ring laser gyroscopes, have introduced them in the Boeing 757 and 767 and Airbus A 310, now in production. These gyroscopes are ideally suitable for the various guided missile applications the Defense sector also.

Air Reconnaissance

Lasers can be used as a secretive illuminators for aerial reconnaissance during night with high precision. Earlier it was done using a camera, equipped with either magnetic flares or powerful strobe lights with their cumbersome power supplies. For this purpose, a helium-neon laser or a gallium arsenide semiconductor laser is used. Two properties of the laser, namely, its narrow beam and its radiance or brilliance are of importance in this particular application.

The block diagram of the laser camera is shown in figure. One of the beams passes downwards through a six-sided prism scanner towards the earth. The prism scans through a selected angle at right angles to the direction of the flight of the aircraft. The other beam passes through a Pockels cell modulator. On emerging from the modulator, the beam strikes the prism scanner and is then reflected towards and recorded on the film.

The laser beams reflected from the target area are picked up by a Schmidt lens, which images the light on to a photodetector. The video output of the photodetector, corresponding to the reflectivity of the observed terrain, drives the modulator. Thus, the returned beam modulates the original beam. The pictures thus obtained are comparable in resolution with those taken under daylight conditions. Thus, the enemy targets can be photographed at night under high secrecy during the flight of the aircraft.

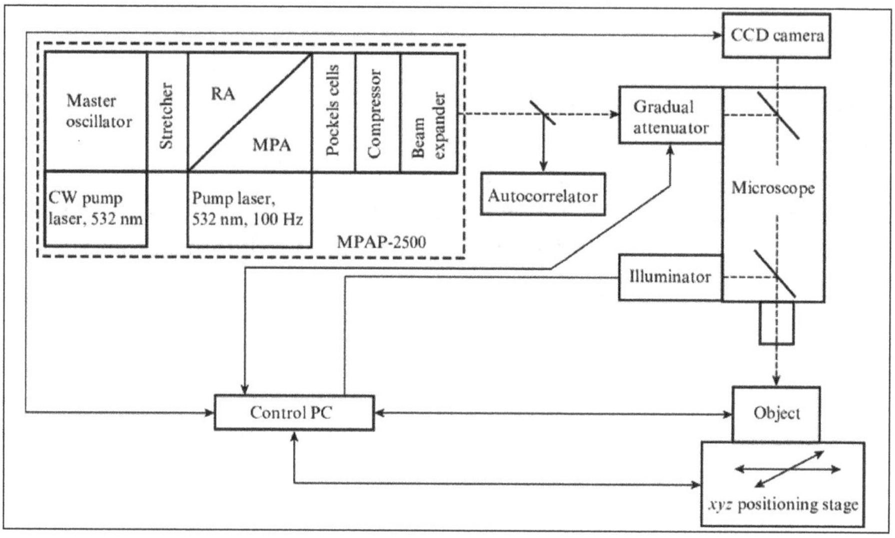

Communications

A very useful and interesting application of laser is in the field of communications, which takes advantage of its wide bandwidth and narrow beam width over long distances. The laser beams can be created in a range of wavelengths from the ultraviolet to the infrared regions of the electromagnetic spectrum. The colour of the emitted light is relatively not important. The infrared region is preferred by the military, as it is more difficult to detect. The advent of semiconductor lasers has made possible the use of lasers for signal transmission. They are excited directly by electric cur-rent to yield a laser beam in the invisible infrared region.

A particular aspect of laser transmission, which makes it preferable to the ordinary radio waves for military purposes is the strict secrecy provided by the narrow beam width. Since no unwanted reception outside the narrow bundles of rays is possible, a high degree of secrecy can be maintained between two points, and thus, an interception-proof communication network can be realised. Besides, laser communication system is immune from jamming and from interference by spurious radio noise.

The optical laser has a great potential for use in long distance communication. Since the capacity of a communication channel is proportional to the frequency band width, at optical frequencies, the information carrying capacity is many times more than that is possible at lower frequencies. This and the fact that the laser is a generator of highly coherent beams which are powerful and sharply directed, make it ideally suited for communications.

In this regard, microwave technique offers direct competition to the laser as it has been perfected already to a high degree. Moreover, the optical frequency waves suffer a considerable disadvantage in case of atmospheric transmission since they are attenuated considerably by snow fog, and rain. Therefore, the laser communication through the atmospheric medium is effective only in clear weather conditions, with no obstacles interrupting the beam between the transmitting and the receiving stations.

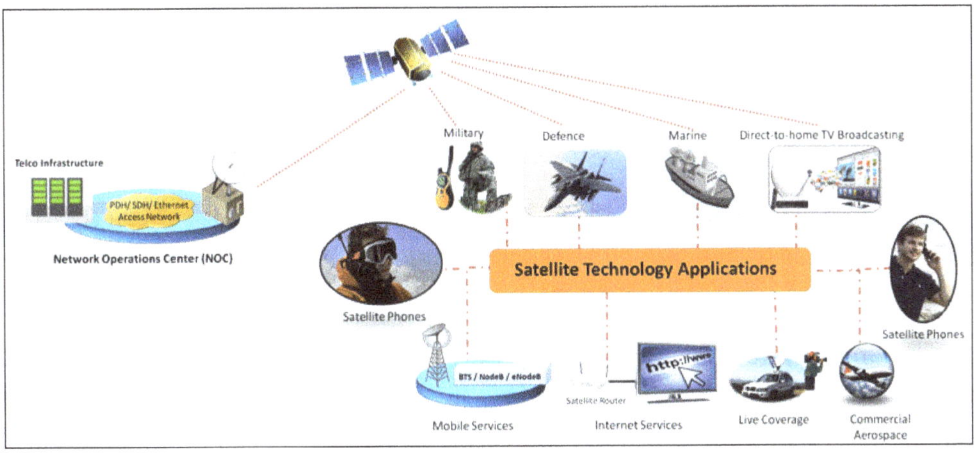

Figure shows the principle behind the long distance communication system, which involves multiplexing the simultaneous transmission of different messages over the same pathway. For example, a channel for transmitting an individual human voice requires a frequency band extending from 2000 to 4000 cycles per second. For modulation of the signal without the addition of any noise, the carrier wave should be of a very narrow spectral width. This single frequency carrier wave is then successively modulated by a large number of voice signals to create a new composite single wave. With the help of special electrical networks, several broad communication bands are combined for simultaneous transmission over a single intensity pathway. On the other side of the line, a similar network separates the single signal into its component broad bands which are demodulated into individual telephone calls.

Thus considerable economy and efficiency in communication is achieved through multiplexing process. Since an individual communication channel requires the same bandwidth regardless of the region of the spectrum in which it is located, it is quite obvious from the above that the visible and near-infrared laser frequencies, which are about 1,000,000 times the frequency of the millimetre waves, offer great economy for communication.

For communication purposes, the laser beam is modulated by the signal. At the receiving station, the modulated beam is demodulated (detected) to separate the required signal from the laser beam (carrier). The output current, which varies with the intensity of the signal, is amplified and then fed to the speaker.

Most of the optical modulators devised so far are based on the variations in the refractive index of the substance used according to the signal wave. The continuous laser output from a laser passes through a polarisation modulator (KDP crystal) as shown in figure. Ring electrodes are placed on the crystal and an electric field proportional to the signal wave is applied to the crystal, parallel to the axis. Due to the change in refractive index of the crystal, which follows the electric signal, there is change in polarisation of the light beam. As a result of this, the intensity of the light coming out of the analyser also changes, according to the signal.

The laser beam from a semiconductor laser can be directly modulated by varying the current through it, according to the signal. The demodulation of the laser beam can be accomplished in two ways: (i) by direct photodetectors and (ii) by photomixers. Photomultiplier detectors are good to use in the visible and infrared regions. The method of demodulation by photoelectric detectors is shown in figure. Silicon photodiodes, developed rapidly after the discovery of the laser, have a peak response at about 8500- 9000 Å (one Å= lxlo-8 cm). This being the spectral region of the gallium arsenide lasers, the silicon photodiodes can be used as sensitive detectors in that region.

Demodulation by optical heterodyne detection is done by superposing on the incoming incident signal, a beam of light from an unmodulated laser called a local oscillator and allowing the resulting combined beam to fall on a photoemissive surface of the photodetector as shown in figure. The electron current from the detector is modulated at a frequency equal to the difference between the signal and the local oscillator frequencies. When this is fed to an audio speaker, the input signal of the communication is reproduced.

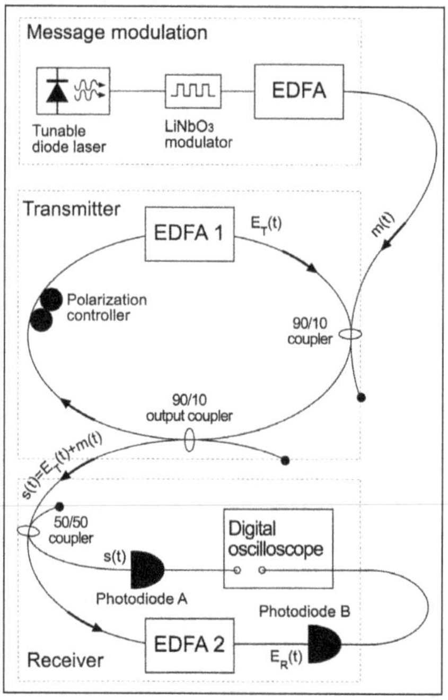

Laser communication through open atmosphere is possible only when there is line of sight between the transmitter and the receiver and that too in good weather conditions. To circumvent these difficulties, laser communication through the medium of optical fibre has been achieved in recent years.

In an antimissile defence system, laser is used to dispose the energy of warhead, not by vaporising or melting it, but by partially damaging the missile, say by drilling a hole.

Tremendous energy is required to completely burn the missile, which is not practicable. If a guided vane of a missile is fractured, several vibrations will be developed in the air frame thereby disintegrating major sensitive portion of the missile.

Two types of anti-missile defence systems have been visualised. One such system, laser kill system is completely earthbound. Here, an early warning microwave radar gives a rough position of the approaching missile. Then a lidar aligned to the target by the tracking radar gives the precise position of the missile. This data is fed on to another high intensity laser beam which actually does the killing. To exploit the laser's killing capability, a high speed servo system and a complex focusing system are essential.

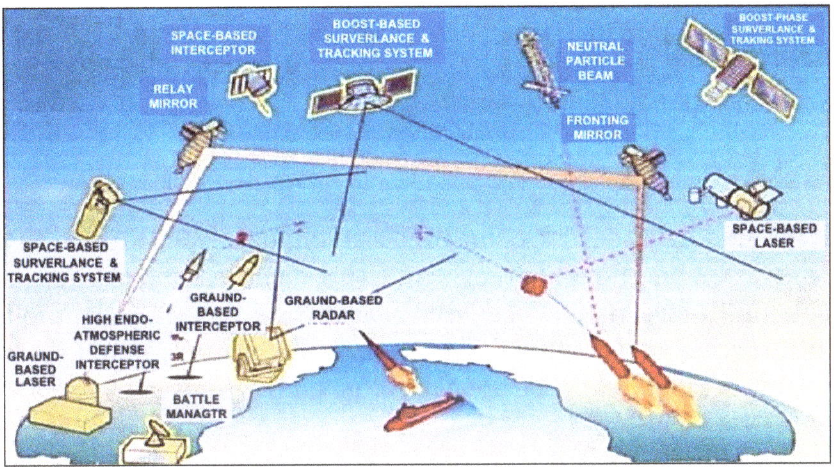

The other anti-missile defence system is the orbiting space station, equipped with detecting, tracking and killing laser devices. An infrared homing system on the laser weapon is used to close on an enemy vehicle and then fire a high energy laser beam. Firing by laser weapons would not change the positional or altitude stability of the space station. It is predicted that the lasers would ultimately make inter-continental ballistic missiles (ICBM's) obsolete.

There are, however, many limitations in the utilisation of laser in its anti-missile role. The power required is very prohibitive and as a result huge power stations are required for the operation. At present, huge power of more than 100 kW in the continuous mode is being obtained from the gas dynamic carbon dioxide lasers and some other chemical lasers, developed in the US and Russia. This amount of laser power is sufficient to destroy an enemyvehicleora missile. The SDI or Star Wars is the US programme aimed at defending itself and its allies against strikes through space. The concept of strategic defence is concentrated at a three-layered defence system in which the enemy missile can be destroyed in the boost, or the mid-course, or terminal phase. Each layer of defence employs several alternatives of weapon systems. Laser is one such important system.

For detecting and destroying missiles, different types of high power lasers, such as gas dynamic carbon dioxide, excimer, x-ray, free-electron, and chemical lasers can be used.

In one such programme, the us scientists are developing 5 to 10 MW deuterium fluoride lasers for destroying the ICBM. In the ground-based laser systems, laser beams will be directed towards a large mirror in geosynchronous orbit. From there, the beams will be directed towards a moving mirror in low orbit which will reflect the beams on to the missile to destroy it. The nuclear-pumped x-ray lasers are also being considered for destroying missiles in the boost phase.

Laser Proximity Fuze

The proximity fuze, developed in the US using a solid-state laser, detonates the missile warhead when it comes within the range of its target. The higher manoeuvrability of the missile is expected to improve its performance a great deal in close in aerial combat. It is also claimed that the proximity fuze and the warhead will enable the missile to destroy its target without hitting it directly.

Laser Beacon

The present infrared light sources, being used as ground beacons to identify the ground points, are inefficient and not much reliable. Using a lensless diode array, a laser beacon can made multi-directional. The laser beacons are light in weight, efficient and have long life. Another advantage is that the pulses can be used ground beacons so that the air- dropping of supplies can be done at the given locations efficiently.

Weapon Firing Simulator

If we take into account the cost of ammunition and large land that is required to fire it, basic training of the tank gunners is very expensive. A simulator technique, which does not sacrifice the acquisition of the basic skills during trials, has been developed in which the main weapon has been replaced by a laser. The technique is known as the weapon firing simulator.

The simulator, installed in the machine gun mount of the vehicle, produces a single burst intense red light when the firing circuit of main weapon is activated. On seeing this bright spot of light, which is visible momentarily in optical system of the vehicle, the trainee is able to lay his gun sights accurately to track the targets. It also enables one to check the accuracy and proficiency of the crew in operating the vehicle's main weapons system.

Laser Applications in Civil

Laser Drilling

Laser drilling of metals is based on a face-heating phenomenon. The absorbed intensity is transformed into heat within the penetration depth of laser radiation. And when the illuminated spot on the surface reaches boiling temperature, material removal starts due the processes of vaporisation and melt expulsion. Laser enables drilling of a diamond die in a few minutes as against 20 hours taken by conventional methods. There is no wastage in the process and the saving in terms of the cost of diamond dust helps in recovering the financial outlay on such a drilling system.

Laser light energy is primarily applied in effecting micro openings in rubies and diamonds. Without heating the entire machined unit, now possible to drill filament canals in refractory materials. Laser drilling of holes in a diamond takes 2 to 3 minute as against 2 to 3 days taken by conventional drilling. A laser installation in Russia drills holes with diameters and depths from 0.005 to 0.8 mm and depths from 1 to 3 mm, respectively, in diamonds of any size and shape. The plus point about laser drilling is that it does not cause any damage to the diamond or any other processed material.

For laser drilling, usually pulsed carbon dioxide, *Nd:YAG* or alexandrite laser is used. Special operations like drilling of holes with diameter less than 0.5 mm using conventional techniques are difficult. However, the pulsed laser microdrilling is quite successful for such operations, both for metals and non-metals. The *Nd:YAG* laser emits at 1.06 μm and the alexandrite laser is tuned at 755 nm. For some metals, the alexandrite laser consistently gives a cleaner entrance hole with more roundness and finer edges than is obtained with Nd:YAG laser. The superior performance of alexandrite laser is due to its shorter wavelength and continuous spiking of its output. The absorption of most materials increases with shorter wavelengths, giving alexandrite laser an edge over the Nd:YAG or laser in many material processing applications.

Laser Micromachining

In the recent years, lasers have also found applications in the field of microelectronics where the laser beam interacts directly with circuit boards and semiconductor chips containing memory and logic circuits. Such laser processing applications have made possible due to development of lasers with increasing stability, high repetition rates, wavelengths well into ultraviolet region, and short pulse durations. These features make it possible to heat discrete micrometer-sized regions reliably and repeatedly to very high temperatures. When applied to local regions of a silicon chip, laser heating can occur without producing damage to neighbouring material or adjacent circuitry. Laser machining is capable of forming tiny electronic circuit patterns directly on to ceramic substrates in one step. The process makes use of a laser assisted by a computer so that it is programmed to describe type of circuit pattern to be machined.

Modern lasers and their associated automatic control equipment are being used for trimming the electrical circuits. The techniques used to switch and move the laser beam are automatically controlled, every operation is designed with object of saving time. For example, it can be arranged to test parts of the circuit while the beam position is being changed. Faster trimming operations can also be achieved by using laser at a higher power.

Laser Cutting

Continuous wave lasers like carbon dioxide gas lasers are extensively used for cutting a wide range of materials, such as graphite, diamond, tungsten, carbide, all metallic foils, ceramics, sapphire, and ferrite. In most cases, continuous cutting is carried out with assist gases like oxygen, carbon dioxide, or air, which produces both mechanical and chemical action intensifying the thermal effects. This gas-assisted cutting is applicable to the metals of thickness up to 5 mm with cut-widths down to 30 μm. The most promising field of laser cutting is the cutting of steels of small thickness (several millimeters) and also of non-metallic materials.

Use of laser cutters in the garment industry, a new and very useful application of the lasers, has been introduced recently in the developed countries. With the aid of computers, lasers can cut clothing many times faster than the tailors using old techniques. It is now possible to slice the through several layers of thick cloth accurately and in a short time using a laser cutter. The laser system also consists of a computer, programmed with cutting instructions and patterns for various to garments. The laser beam, focused on the material cuts through the fabric, leaving impeccably he smooth edges.

Laser Hardening

The principle of laser hardening is the irradiation of material surface for a short time. Heat is conducted into a metal causing quick heating of a thin layer. During the heating period, a high temperature gradient is built up in the surface zone followed by a rapid self-quenching by the cool sub-surface material. The complete hardening d cycle takes about 1 to 2 s. The conventional surface hardening processes, such as flame hardening or induction hardening, often cause a amount of distortion so that the work I becomes waste or there are high additional costs.

Medium and high carbon plain steels more than 0.3 per cent carbon are generally well suited to laser hardening. Their maximum hardness is a function of their carbon content. mount of alloying elements, such as chromium manganese or molybdenum increases the hardenability. The advantage of laser hardening of steels is the possible substitution for expensive alloy steels. With the development of new techniques, lasers are now also being used to harden several other industrial products.

Normally for hardening a track width few millimeters, an output of more than 1 kW from the laser is considered desirable. Recent many laboratories engaged in laser application and research work all over the world, multi watt lasers have been installed. For heat treatment and hardening, usually a continuous wave beam from a carbon dioxide laser is used. To avoid accidents, safety precautions like use of laser goggles are essential.

Metrology

With its high degree of coherence monochromaticity, laser is ideally suited in metrology-the science and system of measurements for measuring lengths, velocities and optical characteristics of various media. An accurate method of measuring distances is in terms wavelength of light. A conventional technique uses an instrument called interferometer for optical measurements.

Using a helium-neon gas laser as the light source in the interferometer, the highest accuracies are possible in the various optical measurements. In fact, the international standard of length, i.e., one meter is calibrated with the help of a frequency-stabilised laser. In the Apollo II mission, American astronomers aimed laser beam on the moon and determined the distance of the moon from the earth to an accuracy within one fool This type of refinement in optical measurements will enable scientists to measure precisely the factors like lunar orbit motion, lunar radius, and fluctuations in the earth's rotation rate. Ultimately, it will be possible to say definitely, whether the continents on the earth are drifting and if so to what extent.With the help of a laser altimeter, the precise height of an aircraft at a given time and also the rates of its ascent and descent can be determined with high accuracy.

The ordinary altimeter depends on air-pressure changes for its readings and thus is not dependable at low-flying levels. The radio altimeter (a radar which reflects radio microwaves received from the ground) has limited accuracy for short distances. Laser altimeter, working on the principle of laser range finder, sends short laser pulses towards the ground from the aircraft and the reflected pulses are detected. Time taken by the laser pulses to travel to ground and back is measured and expressed in terms of height. The device has been used to measure heights up: 3000 m from an aircraft with an accuracy of 1 m cloud altitude meter with a laser is being used to find the range of the cloud accurately. It also works on the principle of a range finder. The light pulse from a laser is sent upwards in the air and the time taken for the pulse to hit the cloud and its detection after reflection, gives the distance of the cloud. The range has to be determined in both clear and bad weather conditions like haze or fog. Since the cloud particles have size of about a micron, the scattered intensity would be more for visible wavelengths. Hence, ruby laser emitting powerful red pulses is highly suitable for this purpose. With this system, it possible to determine the height of the clouds up to 10 km with an accuracy of ±5 m.

Laser for Surveying

A breakthrough in surveying has been achieved by using laser light with geodimeter, a surveying instrument. Previously, a mercury vapour lamp was used to flash a beam of light from one point to another. Lasers have enabled the US coast and geodetic survey teams to map about 5 per cent more terrain in a given time while improving the accuracy to within 1 cm in 10 km.

The spectra-Physics of the US has developed a geodimeter which uses helium-neon gas laser. The instrument called geodilite combines the unique properties of laser light and automatic precision receiver electronics into an instrument capable of greater accuracy, longer range and faster measurements.

Civil Engineering

The negligible divergence of the laser beam stimulated a number of ideas for providing hitherto impossible accuracy and sensitivity in the alignment of tools. Serving as an optical axis, the beam guides the machines used for levelling the concrete facing of the airfields, checking the verticality of the framework of tall buildings, sinking mines, and cutting tunnels from two ends and joining them without tilt.

In Moscow, a laser centring device was used to control the vertical axis during the construction of a TV tower with a precision of 6 mm. In the US, a hard rock boring machine has cut a tunnel, 6.5 m in diameter and 2 1/4 km long, without deviating from its planned course more than 1.58 cm in any direction. This feat was accomplished with the help of the laser beam.

Laser direction finders are widely used in coal mines. In view of the high rate of digging in coal mining, the biggest difficulty is to maintain accurately the given direction, which is achieved using a highly directional laser beam.Similarly, geodolite is used for the detection and measurement of the deformation of large dams and bridges.

Optical Fibre Communication

The communication using light as signal carrier and optical fibres as a transmission medium is termed optical fibre communication. Since the the first commercial installation of a fibre optic system in 1977, the applications of optical fibre communication have increased enormously. Today, every major long distance telecommunication company is spending millions of dollars on optical fibre communication systems. In a optical fibre communication system, voice, or data are converted into a coded pulse of light using a suitable light source. This stream is carried by optical fibres to a regenerating or receiving station. At the final receiving station, the light pulses are converted into electrical signals, decoded, and then converted into the form of original information.

In future, fibre optics is going to be the choice for many communications applications. The biggest advantage of a light wave system is its tremendous information carrying capacity. There are already systems that can carry several thousand simultaneous conversations over a optical fibres, thinner than human hair. In addition to this extremely high capacity, the light guide cables are light weight, immune to electromagnetic interference, and very cheap when compared to copper cables.

Optical fibres used in communication waveguides made of transparent dielectrics whose function is to guide light over long distances. An optical fibre consists of an inner cylinder of glass called the core, surrounded by a cylindrical shell of glass of lower refractive index, called the cladding.

Optical fibres may be classified in terms of the refractive index profile of the core and whether one mode (single mode fibre) or many modes (multimode fibre) are propagating in the fibre. If the core, which is typically made of a high-silica-content glass, has a uniform refractive index, it is called a step-index fibre. If the core has a non-uniform refractive index that gradually decreases from the centre towards the core-clad interface, the fibre is called a graded-index fibre. The cladding surrounding the core has a uniform refractive index that is slightly lower than the refractive index of the core region. If the core diameter is very much reduced so that only one ray can pass through the fibre, it is called a monomode step-index fibre (or single mode fibre). Figure shows a block diagram of an optical fibre communication system.

The basic purpose of an electrical receiver is to detect the received light incident on it and to convert it into an electrical signal containing the information impressed on the light at the transmitter end. An optical receiver consists of a photodetector and an associated amplifier along with necessary filtering and processing. The amplifier converts this current into a usable signal without introducing noise to distort the signal. In the usual fibre optic communication systems, the photodetector used is either a semiconductor pin or avalanche photodiode (APD).

Applications of Optical Fibre Communications

Due to the unique advantages of the optical fibre communications, namely, the tremendous information carrying capacity, freedom from electromagnetic interference, light weight of fibres, longer distance between repeaters, and freedom from signal leakage and crosstalk, this new technology is finding immense applications in the of telecommunications. Already, the telecommunications market for optical fibres has exploded in several developed countries, like USA, UK, France, Denmark, Germany, Japan, etc. The need for greater circuit capacity coupled with the problem of congested duct space led to the initial applications of optical fibre communication for the inter-office trunk in the big cities. Today, the biggest market for optical fibre is the long distance communication. Japan has already completed its long distance routes using single mode fibres. In the US, thousands of kilometres of optical cables are being installed by giant companies, like AT&T, GTE, BELL, etc. Submarine optical fibre cable across the Atlantic Ocean are operational for the last few years.

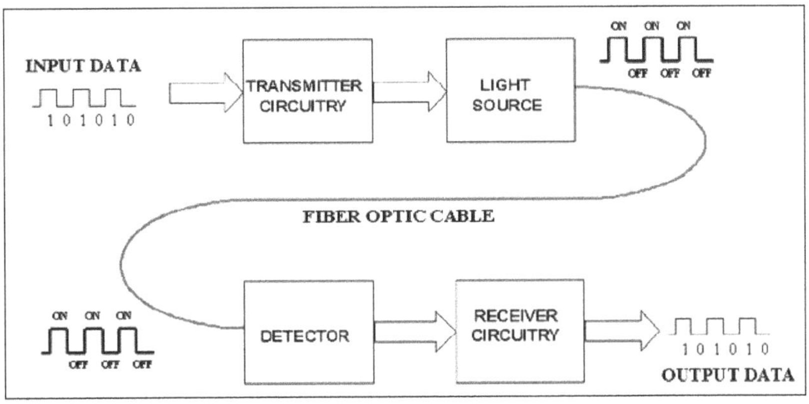

A system with an *AlGaAs* laser for 850 nm wavelength and a graded-index multimode fibre is applied to intra-city networks with bit rates of 32- 140 Mb/s and with transmission spans shorter than 10 km. For distances more than 10 km, a system with in GaAsP laser operating at 1300 nm wavelength and with a graded-index multimode or single mode fibre is preferred in inter-city networks with bit rates higher than 100 Mb/s. When more telecommunication channels are required in a metropolis, the optical fibre telecommunication system is quite effective because the special fibre cables which give more bandwidth can easily replace the existing metal-wire cables in the duct Majority of the present companies use single mode fibres that operate at 1300 nm wavelength with future upgradability to 1500 nm where the attenuation loss will be the least. Already, commercial systems have been installed with bit rates as high as 565 Mb/s which is equivalent to 7680 two-way conversation over a pair of fibres. Since fibres have very high capacity and can transmit voice, data and video, efforts are underway to install fibres into individual houses, with universal information system. Systems are already in use that provide continuing interactive service smoke and heat detectors to automatically alert fire alarms, police alarms and medical alert alarms to summon aid. There is also a potential for completely automating all the control needs of household.

Networks using optical fibres to transmit voice, video and data within a building or within industrial complexes and university campuses have been offered by several vendors. These systems are called local area networks and they cal improve communication inside a high-use area, reduce the bulk of the copper cables, and eliminate congestion in computer rooms.

Optical fibre communications are also(being used for industrial applications, such as process control in nuclear, petrochemical, chemical and food industries and numerical control in large data systems in airways, shipping, railways, gas and oil transportation. Another industrial use of fibre optics is computer applications. Fibres are ideally suited for internal links that require very high data rates of the order of gigabits per second (Gb/s). Auxiliary equipment require lower data rates and hence can be handled both by fibres or copper wires. Fibres offer the added advantage of longer distance network and error-free operation; because their transmission is unaffected by the electromagnetic noise. They will be used in greater volume as inter- and intra-computer links.

Optical fibres are also very useful in Defence applications. Fibres can be used for data links in aircraft with tremendous reduction in weight and increased information capacity. Such aircraft include surveillance and attack aircraft and strategic air command bombers. Similar applications exist for inter- and intra-ship communications, submarine mobile command centres, ship-to-satellite communication links and all types of missile guidance systems. All the above types of applications exist today, making the military applications a large market for fibre optics communications.

Another interesting application is for sensors. In general, a sensor system consists of an electronic control module, sensor head and fibre optic cable. The sensor head senses pressure, temperature, velocity and reaction and converts it into a change in the optical signal which is then analysed to measure the desired change by the electronic control unit.

Lasers have great potential in TV and the cinema. Instead of microwaves as the carrier, laser beam can be used for the transmission of television programmes through optical fibre cables. Since the channel capacity is very high, many TV programmes can be accommodated in a single laser beam.

Data Storage

The storage of higher density of data is possible by using optical techniques. The storage medium is generally a thin film of metal whose optical properties change when it is illuminated with a powerful write laser. The less powerful read laser reads the change in optical property as the required information. Since laser beam can be focused on the spots smaller than one micro diameter, it takes less than one square micro record one bit of information, i.e., 100 million per square cm. Laser video and compact disc are examples of such data storage media in the entertainment market. The magnetic data storage vices like the present day video cassettes in market cannot have such high density data age. However, the main drawback of optical storage is that it is not erasable; such eras optical discs are expected to come into the market within a few years.

Holography

Another important application of laser beam is the production of true three-dimensional pictures in space without the use of lens. The record of' this three-dimensional image of the object on a film is called a hologram. For this pose, the phenomenon called interfere produced due to interaction of two beam monochromatic light waves under certain conditions, is used. The interference pattern is produced by mixing two beams of the laser light. laser light is split into two beams. One beam called the reference beam, is directed toward photographic plate. The other beam is dire towards the object to be recorded, so that sl1itably reflected from the object. The refer. beam and the reflected beam are made to with each other to form an interference patter the photographic plate.

To get a high quality hologram, radiation of in high coherence is required. Generally, the helium-neon gas laser which gives highly coherent and Its I continuous output beam, is used for this purpose. Sometimes, high power pulses from a solid-state laser like ruby laser, are also used. The exposure time, in the case of continuous gas lasers may be from several seconds to a few minutes whereas a giant pulsed laser requires only a few nano- seconds. The holograms are recorded on special le photographic plates with good resolution. The plate is developed and fixed in the usual way.

There is a kind of magic about the hologram. With naked eye, it does not give any picture of the object but only an exposed negative with grayish random patterns on the surface. When it is positioned in its holder and viewed in a monochromatic light beam, such as that produced by laser, the random patterns are transformed into le a sharp and detailed three-dimensional image of b- the object. The image looks like a solid object hanging in space in three dimensions; the viewer may change the angle of view by moving his head from side to side or up and down.

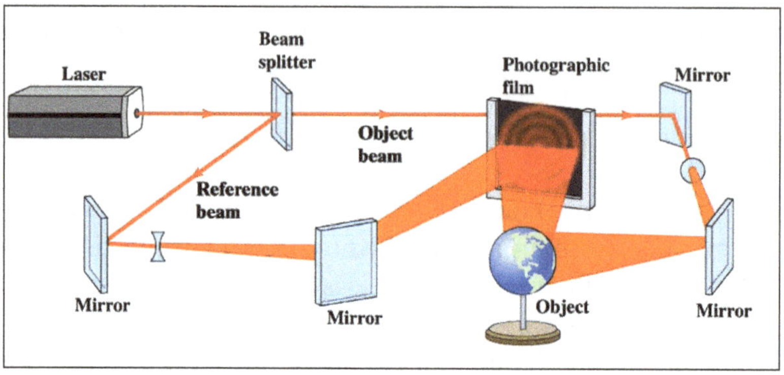

Holography is being used for non-destructive testing, holographic information storage, display devices and pattern matching procedures for 'a such tasks as credit card and identity card verification. Holographic methods can also be used for is secret communication of information by recording e the holograms of secret documents, maps and objects, and restructuring the images only at the receiver end. Interference holography can be used to measure accurately how structures deform under the effect of mechanical stress or thermal gradient.

Standard holograms may be used in industrial production processes to check high precision components with regard to their shape dimensional accuracy. First, a hologram is from a masterpiece according to the method shown in figure. The reconstructed image (hologram serves as a kind of reference. The to be inspected (the deviations of which are checked), is placed at the spot where the master was earlier placed. If the specimen is now exposed in the same way as the master and watched through the developed hologram, small deviations of shape with reference to the original are indicated by interference lines on the specimen. a study of the sensitivity of the interference method (any deformations of the laser light result in an Interference line), even the mistakes within the component (e.g. cracks, cavities, etc.) are indicated in the interference link patterns.

Some day, holographic technique may even be used for target recognition from air to ground and we may eventually have holographic movies and television, giving better appreciation of the objects projected to the viewers.

Medical Applications of Lasers

Lasers are extensively used in medicine and surgery. The first practical application was in eye surgery, where laser was used to weld detached retina and photocoagulate the blood vessels thatgrow into the region in front of the retina, thereby blocking vision.

The laser beam easily passes through transparent portions of the eye, including Cornea and lens to the region of its intended use where its energy is absorbed for treatment. Retina is a sensitive membrane inside the eyeball. Its detachment from the surrounding choroid coat initiates due to a hole in the retina caused by an injury or degenerative changes during the old age. This makes the thick fluid vitreous humour seep and fill itself between the retina and chord oat. The pressure between the retina and choroid coating damages the retina which Soon gets detached from the optic nerve at the back of the to cause blindness.

Before the invention of laser, this delicate operation was done by irradiation of the eye with a xenon arc lamp or even by focusing the sunlight on to the choroid coat. This method involved exposure time to concentrate sufficient he the site of the detached retina. The process cumbersome, painful and relatively slow. Be the patient had to be anaesthesised to prevail eye from moving. Using a high energy pulsed laser, like Nd:YAG, the intense laser light focused as a tiny spot at the detached retina 'welding' it to the underlying choroid coat of the a short time (of the order of one-thousandth of a second). The operation is painless and doe affect the surrounding healthy tissues. Laser also be used to burn out small tumours on the surface of the eye and also those in the vessels of the eye. It is being used to treat coma, cataract, sealing of the retina and even viral diseases of the eye.

The laser cane which is a boon for blind persoiiS operates on the principle of a radar. Two lasers within the cane provide pulses of infrared light which are reflected from points, a short distance in front of the cane. Each reflected beam returns to a photocell inside the cane. The two photocells activate pins in the handle. When the path is smooth, the two pins vibrate steadily. Any hole or other obstacle scatters the light from at- least one of the lasers and stops the vibrations, thus warning the user. The device operates on four small batteries which last up to ten hours. It allows a blind person to scan the area ahead of him and have an idea of the object's shape, distance and dimensions by variations in pitch and intensity of the tone it emits.

Lasers are increasingly being used for the treatment of many different types of cancer. A laser is less damaging than x-ray therapy and surgery; and in many cases, it is quite effective. The use of lasers to remove certain forms of cancerous growth in the body has heralded an era of knifeless and bloodless surgery. It is very effective in curing the diseases of gynaecology, ear, nose, throat, tongue, palate, and cheeks. It is curative in most early cancers, and in late cancers, it is useful in reducing the tumours to facilitate surgery.

Photodynamic therapy (POT), a new exciting form of cancer treatment, combines laser with light-sensitive dye, hematoporphyrin derivative (HPD). This substance, derived from the cow's blood, travels throughout the body of the patient and settles in the malignant tissues. A red light from argon pumped dye laser, focused on the area activates HPD, and the energized substance releases a highly reactive chemical that destroys the cancer cells. Reports indicate that POT is 80 to 90 per cent successful in causing total or neartotal regression of tumours, even after all other forms of therapy have failed. It is highly selective for a diseased tissue, leaving healthy cells relatively untouched.

At some medical centres in the US,searchers have used laser to treat colonic and other types of gastrointestinal cancer. Using endoscope, the laser energy is used to destroy neo- plastic tissue while preserving bowel wall integrity. In some cases, rectal polyps were removed using the CW argon laser, delivered with a power of 4-5 W.

With the development of optical fibres lasers are being used for heart surgery. A common problem with the arteries is the build up of plaque on their interior walls. The plaque, consisting fatty material, calcium, etc, blocks the coronary. arteries reducing the blood flow through the This results in Angina pectoris, a condition that afflicts millions of people worldwide. If the coronary artery is partially blocked, the situation can sometimes be improved by using a method called angioplasty. When substantial blockage of the coronary artery is observed, a laser beam se through the optical fibre could be used to vapourise the plaque, opening a clear channel for smooth flow of blood. This method is called laser angioplasty or vascular recanalisation. Usually argon-ion, Nd:YAG, and carbon dioxide lasers al used for this purpose.

Another important use of the fibre-optic laser catheter is in the treatment of bleeding ulcers. The laser light can photocoagulate blood, thereby causing the cession of bleeding. For this purpose among the three important lasers(carbon dioxide, Nd:YAG and argon- ion), the Nd:YAG laser is preferred because it penetrates deep into the tissue and its effects are not localised at the surface. Using a laser endoscope, small tumours in the urinary bladder are destroyed. Similarly, Nd:YAG and dye lasers are also used to rapidly heat and shatter urinary stones in the kidney.

Laser can also be used for dental treatment Laser beam is useful for charring tooth decay through a painless process called laser glazing. The beam from a high repetition pulsed laser can be focused on dark decayed areas of teeth cavities to destroy the infection in the affected areas in a fraction of a second.

Laser Applications in Surgery

In modern medicine, lasers are increasingly utilized for treatment of a variety of pathologies as interest in less invasive treatment modalities intensifies. The physics behind lasers allows the same basic principles to be applied to a multitude of tissue types using slight modifications of the system. Multiple laser systems have been studied within each field of medicine. The term “laser” was combined with “surgery,” “ablation,” “lithotripsy,” “cancer treatment,” “tumor ablation,” “dermatology,” “skin rejuvenation,” “lipolysis,” “cardiology,” “atrial fibrillation (AF),” and “epilepsy” during separate searches.

In 1900, Max Planck discovered that light is released, transferred, and absorbed in specific amounts of energy called quanta, and that this was related to the frequency of the radiation and what he discovered to be Planck's constant. Shortly after, Einstein published his work on quantum theory, suggesting that most atoms exist in the ground-energy state (E_0). These E_0 molecules can then be converted to higher energy levels when energy is added to them, and in returning to their ground state, the energy is released spontaneously as photons or electromagnetic (EM) waves. He also discovered that when a photon of the same wavelength collides with an excited atom, the two photons are released concurrently, and therefore have equal frequencies. This idea of "stimulated emission" was years later used in the creation of lasers.

Theodore Maiman ultimately created the first "laser" (light amplification by stimulated emission of radiation) by using an electrical source to energize a solid ruby. Following this bookmark invention, its many possible indications in medicine were rapidly recognized. As the CO_2 laser was known to emit a concentrated ray of light that was easily absorbed by water, it became used to vaporize tissue. The neodymium:yttrium-aluminum-garnet (Nd:YAG) laser created coagulative necrosis within tissue, and the visible light lasers were useful for achieving hemostasis. Over time, several different active media have been used to create new lasers, resulting in their utility in a wide range of medical subspecialties. The goal of this review is to provide an overview of the physics behind laser systems, demonstrating how the same basic principles can be applied to various tissue types to accomplish the desired effect, and how this has led to the wide range of clinical applications of lasers.

Table: The various lasers commonly used in medicine along with the wavelength at which they operate, their absorption chromophores, and their clinical applications.

Laser	Wavelength (nm)	Absorption chromophore	Application
Ruby	694	Pigment, hemoglobin	Dermatology, tattoo removal
Nd:YAG	1,064	Pigment, proteins	Wide applications
Er:YAG	2,940	Water	Surgery
Diode	630–980	Pigment, water (range)	LLLT, PDT, surgery
Argon	350–514	Pigment, hemoglobin	Surgery, PDT, ophthalmology, dermatology
CO_2	10,600	Water	Surgery
Pumped-dye	504–690	Pigment	PDT, dermatology

Laser Physics

A simple laser consists of a laser medium (which determines the wavelength of the system) enclosed between two parallel mirrors, one of which is partially reflecting and partially transmitting. The medium is excited by an electrical source until the number of atoms in the excited state is greater than the number in the ground state (population inversion).

When the laser medium is activated, it begins to release excited photons spontaneously in all directions. However, a small subset of these photons travels along the centerline of the laser system in a unified fashion between the mirrors. The mirrors then reflect these photons and the process of stimulated emission is amplified. The partially transmitting mirror then allows a powerful, cohesive beam of photons to be released as laser light.

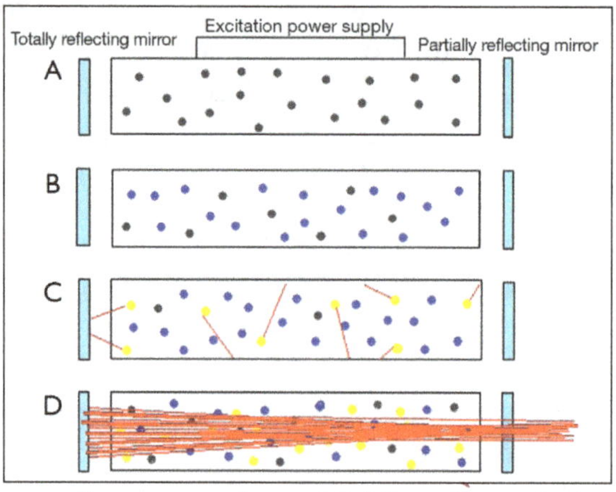

Demonstrates a laser medium at ground state (A) followed by excitation of atoms to higher energy levels (B) and progression to stimulated emission (C) with laser beam generation as a final product (D).

Laser-tissue Interaction

The effect that a laser has on a sample of tissue is dependent on both properties of the tissue as well as the laser. The tissue properties include its structure, water content, thermal conductivity, heat capacity, density, and its ability to absorb, scatter, or reflect the emitted energy. The properties of the laser that play a role are its power, density, energy content, and wavelength.

The main biological targets that are dealt with absorb light very differently, and their optimum absorption spectra depend on the wavelength of the incident photon energy. For the visible light and some near-infrared lasers, the main target chromophores (any substance that absorbs light) are hemoglobin and melanin, whereas for CO_2 lasers, the only chromophore is water. In order to achieve selective photothermolysis (using energy at high peak powers and short pulse widths to destroy the intended target alone) without damaging the surrounding tissue, the target tissue must contain chromophores that absorb a specific laser wavelength, and these chromophores should not be found in the surrounding tissue.

The CO_2, Nd:YAG, and Argon lasers are the lasers most commonly used in medicine and surgery. The CO_2 laser has carbon dioxide gas as its medium and emits energy at 10,600 nm. Because its chromophore, water, exists everywhere, CO_2 lasers cannot be used for selective photothermolysis, though they are tissue-selective. All of the incident energy is absorbed in the tissue water down to a specific depth, preventing deeper tissue damage. CO_2 lasers operate in the invisible infrared waveband, thus requiring an aiming beam for accurate treatment. Focusing the laser on the tissue produces extremely high power density resulting in instant vaporization and ablation of the tissue. As the irradiance of the laser beam is proportional to the inverse of the square of the diameter of the beam, by defocusing the beam, the surgeon is able to easily change the laser from incision mode to bulk vaporization or coagulation. The CO_2 laser has a number of beam modes, each of which reacts differently with the tissue. The simplest mode is continuous wave (CW), in which the laser beam is emitted, operated for a specific time, and then turned off. More recent lasers however are quasi-CW (ultrapulsing), meaning they produce short high-peak power pulses with very long inter-pulse intervals. This has the advantage of allowing more precise incisions with minimal heat build-up because each pulse that is delivered is shorter than the time it takes for the target tissue to cool.

The active medium of the Nd:YAG laser is a single YAG crystal bar covered with neodymium ions. The wavelength of light that is produced by of this system, which is determined by the neodymium ions, is 1,060 nm. Because there are no key tissue chromophores at this wavelength, the Nd:YAG laser-tissue interaction produces largely a scattering effect. Scattering leads to reflection, which prevents the typical narrow, cohesive beam from being produced. This decreases the penetrative ability of the laser, resulting in slower heating of the tissue. This property of the Nd:YAG laser makes it ideal for hemostasis and tumor necrosis, as well as numerous endoscopic procedures within various specialties.

Ion lasers, such as the argon and krypton laser, operate similarly to gas lasers, except they ionize the active medium. This excites ions instead of atoms, using a large power supply. They can operate at both pulsed and CW modes and can produce wavelengths anywhere between 250 and 530 nm, with the two most powerful beams being in the blue (488 nm) and green (514.5 nm) ranges of the spectrum.

Clinical Applications of Lasers

As minimally invasive techniques are continually being sought out for the treatment of different pathologic processes, the use of lasers has become increasingly popular in modern medicine. In addition to their practical usefulness in the operating room, lasers have a wide range of applications in ophthalmology, lithotripsy, the diagnosis and treatment of various cancers, as well as dermatologic and cosmetic procedures.

Lithotripsy

Laser lithotripsy has been a widely accepted technique for the fragmentation of urinary and biliary stones for the past few decades. Lasers can accomplish lithotripsy by having a photoacoustical/photomechanical effect (laser-induced shockwave lithotripsy) or a predominantly photothermal effect. Of the lasers commonly used in lithotripsy, the 1-µsec pulsed-dye laser is the most popular shockwave laser and has been extensively studied (12-14). This device is based on the excitation of coumarin dye to produce the monochromatic light that fragments the calculi. At 504 nm, a green light that is absorbed largely by the yellow-colored urinary calculi is produced, which allows it to be safely used without causing much damage to surrounding tissues. As the stone absorbs the energy from the laser, the excited ions that are released form a quickly expanding and pulsating cloud around the stone, creating a shock wave that then breaks the calculus into fragments. Because this laser is ineffective against the nonabsorbent colorless calculi such as those composed of cystine, photosensitizers (dye) have successfully been used as irrigation fluids and absorbents to initiate the process of fragmentation. The Q-switched Nd:YAG laser also accomplishes lithotripsy by this mechanism, but it generates larger-magnitude of shockwaves.

The long-pulsed Holium:YAG laser on the other hand, uses a mainly photothermal mechanism to fragment calculi. The laser produces light with a wavelength of 2,100 nm, which is highly absorbable by water. Thus in the appropriate environment, fluid absorbs the energy and is heated as a result. A cloud of vapor is produced, parting the water and allowing the remaining portion of the laser light to directly contact the calculus surface, drilling holes into it and leading to its fragmentation. A study conducted by Cimino *et al.* demonstrated that Ho:YAG laser lithotripsy is a more efficacious endoscopic technique for the treatment of ureteral stones with higher stone fragmentation rates compared to pneumatic lithotripsy, and a review conducted by Teichman concluded that this laser is safe, effective, and works just as well if not better than other modalities, and that it may also be used for biliary stones.

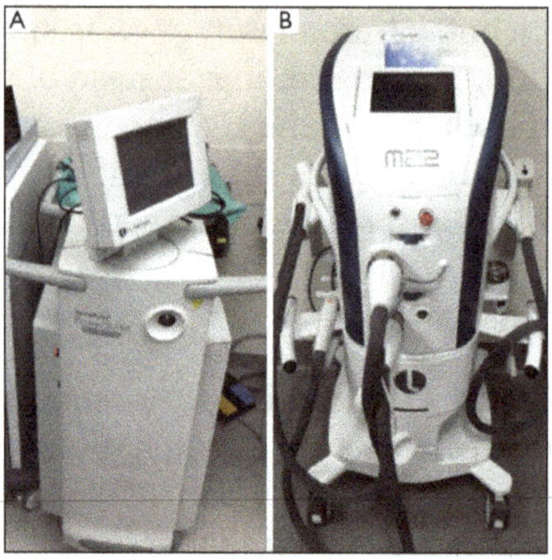

Demonstrates a Ho:YAG lithotripsy laser (A) and an neodymium:yttrium-aluminum-garnet (Nd:YAG) dermatologic laser (B).

Oncology

Lasers are currently being safely used for the treatment of cancers arising in various organ systems. In neurosurgery for example, laser interstitial thermal therapy (LITT) is a preferred treatment option for patients who are not ideal surgical candidates. Since their introduction to neurosurgery, lasers have become increasingly safe to use and have been successfully applied for the treatment of unresectable gliomas as well as hard and hemorrhagic tumors such as meniniomas, tumors of the deep skull base, or tumors deep in the ventricles. Mucosal ablation techniques using lasers are currently being widely and successfully used for the treatment of superficial gastrointestinal cancers including early gastric cancer, superficial esophageal cancer, colorectal adenoma, and high-grade Barrett's esophagus. Moreover, photodynamic therapy (PDT) using lasers has also been shown to be an effective treatment modality for specific types of lung cancer lesions.

Direct laser ablation has been used for direct destruction of cancer cells through its photochemical, photomechanical, and photothermal effects. The photochemical reactions that occur ultimately form toxic radicals that lead to the death of tissues, the photomechanical reactions induce stress on the tissue and lead to its fragmentation and the photothermal reactions induce heating and coagulation, which cause cell death.

To enhance this process and more accurately target the desired tumor cells, PDT was developed nearly a century ago and has gained great popularity since. This treatment modality involves the administration of a photosensitizing drug followed by the subsequent illumination of the target area with visible light corresponding to the absorbance wavelength of the photosensitizing drug. The photosensitizer, which is then activated, initially forms the excited singlet state and then transitions to the triplet state, which in the presence of oxygen form reactive oxygen species that are destructive to neoplastic cells. Selective photothermal therapy, on the other hand, uses localized light-absorbing dye to enhance the laser-induced destruction of the tumor cells.

Aesthetic and Reconstructive Surgery

The unique ability of lasers to target specific structures and layers of tissue makes them a powerful tool in cosmetic and reconstructive surgery. Laser resurfacing has been a major tool used for anti-aging treatment in recent medicine, as the induction of new collagen formation is known to lessen the effects of photoaging. Original skin resurfacing techniques involved using ablative CO_2 and Er:YAG laser systems to target a specific portion of the dermis. However, because these systems also remove a significant amount of epidermis, they result in prolonged recovery and increased side effects such as infections and erythema. Nonablative lasers, such as the intense pulsed light, Nd:YAG, diode, and Er:glass lasers, which mostly release infrared light, were subsequently developed to overcome these issues. The goal of these systems is to target the water in the dermis, which during the process heats collagen and induces remodeling. Because there is a system that simultaneously cools the epidermis, tissue evaporation does not occur and no external wound is produced. Most recently, fractionated laser resurfacing has become the basis of skin resurfacing. Using fractionated lasers, fine beams of high-energy light are used to inducing small zones of thermal damage ("microscopic thermal zones") and treating only fractions of skin at a time.

Laser-assisted lipolysis, which uses an optical fiber inserted inside a 1-mm cannula, has also become an increasingly popular procedure in cosmetic surgery. Due to the small cannula size, a smaller incision is needed, resulting in less bleeding and scar formation.

Of all the lasers that are available for medical purposes, those with 920 nm wavelengths have the smallest absorption coefficient in fat tissue, and so they penetrate the deeper layers of tissue. Those with wavelengths in the 1,320–1,444 nm range have the largest absorption coefficient in fat, causing smaller penetration depth and allowing for superficial treatment of such tissues. The Nd:YAG laser (1,064 nm) is the system that is used most widely in laser lipolysis, as the absorption coefficient of fat tissue in this wavelength results in good penetration depth with medium absorption, causing only moderate temperature elevation and thus less tissue damage. Further, the coagulation of small blood vessels by the laser light at this wavelength results in significantly less blood loss during the procedure. Abdelaal and Aboelatta were able to show a significant decrease in blood loss (54%) with laser-assisted liposuction when compared to traditional methods. Additionally, a review conducted by Mordon and Plot concluded that laser lipolysis produces more even skin results.

Finally, the ability of lasers to selectively target pathologic vasculature makes them an ideal source for the treatment of vascular defects such as port-wine stains. Prior to use of lasers, patients did not have many treatment options for these types of abnormalities. Currently however, lasers that are preferentially absorbed by hemoglobin over melanin are used for this purpose, with little trauma to the epidermis. More recently, lasers with longer wavelengths, and thus the ability to achieve deeper tissue penetration, have also been introduced.

Ablation of Conductive Pathways

After it was discovered that the pulmonary veins (PV) are an important source of ectopic beats that lead to the paroxysms of atrial fibrillation (AF), the development of catheter ablation devices was inspired for circumferential PV isolation (PVI). Today, the laser balloon catheter is one of the endoscopic ablation systems (EAS) commonly used for the treatment of AF. The device consists of a catheter with a compliant balloon at its tip that is continuously flushed with deuterium oxide. The catheter is introduced into the left atrium and an endoscope is then inserted into the catheter shaft, allowing direct visualization of the ablation target inside the heart. Ablation is performed with a 980-nm diode laser that is housed in the central lumen, emitting laser energy perpendicular to the catheter shaft covering an arc of a 30° angle and allowing circular ablation around each PV. Laser at this wavelength is not absorbed by deuterium oxide. As a result, it penetrates tissue beyond the endothelium, where it is absorbed by water molecules, resulting in heating and coagulation necrosis. The energy that is delivered can be titrated by changing the power (5.5–12 W) in a set of predefined levels. The energy levels are altered depending on which cardiac wall is being targeted. The Nd:YAG laser is another laser system that is commonly used for this purpose. A multicenter study conducted by Metzner *et al.* has shown significant success rates of PVI using EAS, and has suggested that the 1-year success rate is comparable to conventional PVI techniques (about 63%).

In order to successfully result in a complete conduction block, a fully transmural lesion must be created in the heart. Melby *et al.* demonstrated that electrical impulses, both paced and AF, could still propagate even through very narrow gaps (≥1 mm) in the ablation line. When comparing the effects of different energy levels, studies have shown that the use of higher energy levels results in higher rates of PVI with lower AF recurrence rates and no compromise of the safety profile.

In neurological surgery, MRI-guided laser-induced thermal therapy (MRgLITT) is commonly used to treat refractory epilepsy, either as a means of ablating the epileptic foci, or as a disconnection tool. MRgLITT combines a diode laser (980-nm) with imaging technology to provide intraoperative information that is necessary for controlling the amount of energy delivered.

Researchers demonstrated the successful use of laser ablation for the management of refractory epilepsy of many different focal origins including mesial temporal lobe epilepsy, cortical dysplasia, post-stroke neocortical focus, encephalocele, periventricular nodular heterotopia, and hypothalamic hamartomas. In addition to resective techniques for epilepsy management, there are disconnective treatment strategies that separate the epileptogenic brain from the nonepileptogenic brain by corpus callosotomy or hemispheretomy. Research demonstrated successful endoscopic disconnection of hypothalamic hamartomas with the use of a robot-assisted thulium-laser and Choudhri *et al.* successfully demonstrated the use of a carbon dioxide laser for corpus callosotomy in children.

Miscellaneous Applications of Laser

Laser Printing

For the last few years, there has been tremendous increase in the use of computers as an aid to the management, processing and dissemination of information. The use of computers in generating bank statements, insurance, telephone and electricity bills as well as publicity brochures advertising mass-produced goods are typical examples of this development. The peripheral device required by the computer for all these applications is the printer. Today, use of computers in large data processing installations places very high demand on printers as regards its speed, character flexibility, and print quality. The conventional impact printers can no longer meet these demands because of their limited speed and character flexibility. In the new generation printers, printing method is based on the principle of electrophotography. Since the light source in such printers is a laser, these devices are called laser printers.

Figure is a schematic diagram of a laser printer. A photoconductor drum at the centre rotates at a constant angular speed. Its surface is treated with a photoconductive coating material like hydrogenated amorphous silicon. In the dark, this photoconductor has a very high electrical resistance which drops when the coating is exposed to light. The surface of the photoconductor drum is electrically charged by means of the charge corotron and the charged layer is then rotated past the write exposure station. Only those locations on the drum surface at which the Information to be printed is to appear are exposed. Two exposure facilities transmitting current computer data and the information to be repeated on each print page (company letterhead, tables, bill forms) are available. As a result of exposure, the surface charge flows to ground inside the drum leaving a latent electrostatic charge pattern. The developer station contains an electrostatically charged toner. The paper transport moves the paper up to the drum and to the transfer station. The paper and toner charges are of opposite polarity so that the toner is attracted by the paper and adheres to it. The photoconducting drum is discharged in the discharge station and the process is repeated.

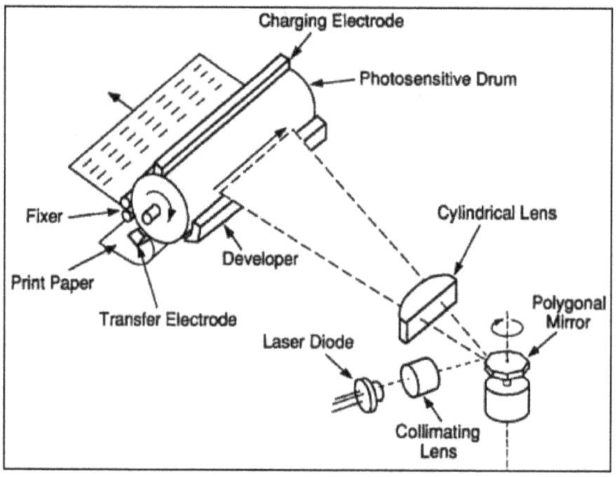

High-performance laser printers offer three important advantages:

- High printing speed, about 10,000 lines per minute at six lines per inch, which makes it possible to print over four million characters in one minute.
- Very high degree of flexibility as regards character generation. Today, the laser printers have character sets of several hundred different characters. Since the print characters are stored electronically, the only consideration limiting the number of characters available in the printer. is the cost I of the necessary memory space.
- Excellent print quality.

Seismography

In its seismographic application, i.e., detection of earthquakes and underground nuclear blasts, the instruments using lasers are ten times more accurate than the conventional devices. This laser application is based on the principle of Doppler shift in the frequency of the light scattered from a moving substance. The scattered beam is mixed with the part of the. incident beam in a detector and the beat frequency is determined, which gives the measure of the movement of the earth's crust.

High-speed Photography

The intense laser light also finds application in high-speed photography for recording extremely fast or transient phenomena like the bullet shot by a gun, armour penetration and the instant of fracturing. Such lightning speed phenomena have been photographed with the help of very short intense light pulses from Q-switched lasers, capable of exposing up to 9,000 frames per second. Ultrashort pulses can be used to study ultrafast phenomena and processes, such as recombination of electron-hole pairs or excitons in semiconductors.

Scientific Research

Lasers have opened new fields of investigation in science and technology. It has given physics a versatile tool for the study of interaction of light and matter. The powerful beam of laser has become an important tool for spectroscopic analysis. A laser system, known as microprobe, is used for exciting emission from solid samples for spectrographic analysis.

CV Raman discovered a new phenomenon, known as Raman Effect by which molecular structures of different substances can be investigated by passing monochromatic light through them. He found that when light passes through a transparent substance, it is scattered and emerges with a change of frequency caused due to the vibration of molecules in the substance. This produces additional lines (known as Raman lines) in the scattered light spectrum. The discovery of laser is a great boon for recording the Raman spectra. The use of lasers has enabled recording of Raman lines within seconds, which otherwise would require long exposure times of or few hours using ordinary light sources. The analysis of Raman lines gives the fundamental properties of the substances.

Similarly, lasers can also be used for analysing liquids. A laser beam, when passed through a liquid, gives several colours (wavelengths) and the process is called fluorescence. The study of the fluorescence spectra thus obtained gives the properties of the liquids.

Lasers offer attractive possibilities in terms of the exploration of molecular structure and determination of nature of chemical reactions. A laser beam can initiate and hasten a chemical reaction. Since different reactions require different wavelengths of light, a 'tunable' laser (i.e., a source whose wavelength can be altered as in radio tuning) is of immense help to a chemist. Tunable lasers, particularly dye lasers, now cover the entire visible spectrum and have revolutionised optical spectroscopy. In photochemistry, lasers with short duration pulses are highly useful for inducing and monitoring ultrafast chemical reactions more efficiently than by any conventional method.

Laser also finds application in biological research. Using laser techniques, biological studies have been carried out in enzymes, proteins, cellular components and isolated cells, microorganisms, tissue culture, isolated physiological systems individual organs, etc. Using a ruby laser coupled with a microscope, single cells have been irradiated with laser beams focused on to a spot of the order of one micron to destroy individual chromosomes, thus making available a highly delicate instrument for genetic studies.

It is also possible to produce laser beams as narrow as the diameter of a protein molecule and use it to alter genetic properties of living organisms.

Environmental Studies

The constituent gases and vapours in the atmosphere can be detected and measured with lasers by means of at least three selective mechanisms. These are: (1) selective absorption of laser light which spectrally matches the natural absorption characteristics of the molecule, resonance or fluorescence scattering of laser light, and Raman scattering.

The absorption lines of water vapour in the atmosphere are very close to the emission wavelength of ruby lasers. By tuning the wavelength of a ruby lidar, one can observe the change in backscatter caused by absorption within the water vapour lines. With the availability of tunable dye lasers, it is easier to take advantage of resonance scattering which helps to identify the constituents of the atmosphere. In recent years, the laser techniques have been well established for the purpose of environmental monitoring, cloud height detection, and urban pollution studies.

Nuclear Fusion

Thermonuclear fusion is the process by which huge energy is produced in the sun and stars. It is the process by which nuclei of light elements such as deuterium (an isotope of hydrogen) are fused (or joined) together to produce heavier elements like helium. In this reaction, a large quantity of energy and neutrons are released.

For thermonuclear fusion to take place, a temperature of about one million degrees centigrade is required. Today, this is achieved by the implosion of the atoms of the material by a focused high energy laser beam. Thermonuclear re. action has several advantages over the fission process. Firstly, the immense energy comes from a very small quantity of material. Secondly, the supply of fusion fuel is virtually inexhaustible as deuterium can be extracted cheaply from the world's oceans. Thirdly, there is no problem of atmospheric pollution. It will be simpler and easier to make a hydrogen bomb which will be 'clean', i.e.,its explosion will be free from the lingering effects of radioactive fallout.

Fire Detection

Laser's application in fire detection is based on the principle that a laser beam is affected by hot gases emanating from a' fire. A focused laser beam is directed across an open space near ceiling level from one side of the room to the other. It is reflected back to a photocell from a mirror fixed on the opposite wall. Any fire starting below this level will cause turbulent hot air to rise. The laser beam, normally steady, is refracted by the temperature gradients in the hot gases and is displaced from its usual position on a photocell. The deflection can be made to trigger an alarm. Results have indicated that the laser beam system is at least as fast as the most sensitive fire detection systems in use worldwide.

Intrusion Alarm

A gallium arsenide diode laser can be used to set up an invisible fence to protect an area. An infrared laser beam (in combination with an optical detector) can seal a path, an area or a volume against infiltrators. When the invisible beam is interrupted by an intruder trying to approach the protected area, it sets off a remote alarm. The laser alarm has many advantages over the conventional electric alarm. The infrared beam, being invisible, cannot be spotted by the intruder. The narrowness of the beam minimised false alarms by the passage of birds, small animals and objects floating in the air.

Ruby, Nd:YAG and Nd:Glass Lasers

Laser rods of ruby, Nd:glass, flash lamps and hard coated laser mirrors, have been developed indigenously at the Defence Science Centre (DSC), Delhi, and the solid sate lasers giving peak power output of a few megawatts have been developed for Defence applications. BARC has also developed these lasers with mainly imported components, Laser range finders with Nd:YAG or Nd:glass as the active element have been developed at Instruments Research Development Establishment(IRDE), Dehradun and DSC. CAT, is developing a high power Nd:glass laser for atomic energy application.

Helium-Neon Laser

Helium-Neon lasers of low power output (2- 5 mW) with lifetimes of a few thousand hours have been developed at IISc, NPL, and Bharat Electronics Ltd., Bangalore. The technology has been transferred by NPL to M/s Laser Instruments, New Delhi and by BARC, Bombay to ECIL, Hyderabad. they started production of these lasers commercially about 20 years back but stopped production since their performance is far from satisfactory. BEL also made an attempt about 10 years back and stopped production due to lack of sufficient technology.

Carbon Dioxide Laser

Carbon dioxide lasers giving an output power in the range 10-100 W have been developed at BARC IIT, Kanpur, IRDE and DSC, Central Electronics Ltd. (CEL) and Jyoti Ltd. have started commercial production of these lasers around 1975 but have stopped production by 1982. CAT has developed transverse carbon dioxide laser with 3.5 kW power.

Semiconductor Laser

BARC and Solid State Physics Laboratory (SPL), Delhi have developed low power gallium-arsenide lasers with a view to use them for applications in communication and ranging.

BARC demonstrated communication over 20 Km distance using laser. Further work is necessary to develop these lasers with heterostructures and to improve their efficiency.

Materials

Basic laser materials like ruby, Nd:phosphate glass and lithium niobate are being developed at DSC for Defence applications, Central Glass and Ceramics Research Institute, Calcutta(CGCRI), has also developed good quality Nd:silicate glass for commercial applications. The development of gallium-arsenide and Nd:YAG crystals is under process at SPL. Several establishments and institutes like DSC, IRDE, IISc, NPL, BARC, IIT, Kanpur and BEL have established optical workshops including coating facilities to fabricate laser components. Good experience has been gained to fabricate laser rods and hard coated laser morrors at DSC.

Fibre-optic Communication

In 1980, a panel on Optical Fibre Communication System constituted by the Electronics Commission recommended the introduction of optical fibre communication in India. With this in view, CGCRI took up an R&D project on indigenous development of optical communication fibre. In1982, a System Appraisal Group for Optical Fibre and Cables comprising representatives of the Department of Electronics, Ministry of Defence, IIT, De/hi, Telecommunication Research Centre (TRC) and Hindustan Cables Ltd. recommended setting up of R&D and manufacturing facilities of optical fibres and cables at HCL through collaborative arrangement. It was decided to manufacture at HCL the multimode graded index fibre with' 3 to 5 db/km loss and bandwidth up to 100 MHz. Similarly, the production of optical fibre has been started at OPTEL, Bhopal with foreign collaboration. In 1983, a Committee on Optical Fibres and Cables (COFC) was constituted by the Ministry of Communication to finalise the technical specifications for optical fibres and cables required not only for communications, but also for Defence and other sectors.

The Department of Communications has Successfully installed the optical fibre cable and an 8 Mb system to provide junctions between two exchanges in the Pune Telephone system. TRC has taken up a design of an indigenous 34 Mb. system to be installed between Thana and Powai in the Bombay Telephone system. Efforts are onie way to introduce optical fibre communication several trunk routes in the country.

Need to Develop Laser Technology

There are large gaps in the development of laser technology and its production between our country and other developed countries. Efforts in this area have been so limited in our country that they are not even equal to the efforts made at one major institution in the USA. Not a single reliable laser system is commercially available in the country. Though some institutions in our country have fabricated some experimental lasers on a laboratory scale, reliable operation of these lasers has still been a problem. As an outcome of status report of SAC to PM, a National laser Programme has been started recently.

Advantages of lasers for various applications in our country are well known and laser research has been recognised as one of the frontier areas to be developed in the 8th Five Year Plan. It is high time for our country to intensify the R&D efforts in the identified areas with time bound pro- Grammies and start the production of lasers for mass applications.

Industrial and Commercial Applications

Laser Cutting

Laser cutting is a technology that uses a laser to cut materials, and is typically used for industrial manufacturing applications, but is also starting to be used by schools, small businesses, and hobbyists. Laser cutting works by directing the output of a high-power laser most commonly through optics. The [laser optics] and CNC (computer numerical control) are used to direct the material or the laser beam generated. A commercial laser for cutting materials involved a motion control system to follow a CNC or G-code of the pattern to be cut onto the material. The focused laser beam is directed at the material, which then either melts, burns, vaporizes away, or is blown away by a jet of gas, leaving an edge with a high-quality surface finish. Industrial laser cutters are used to cut flat-sheet material as well as structural and piping materials.

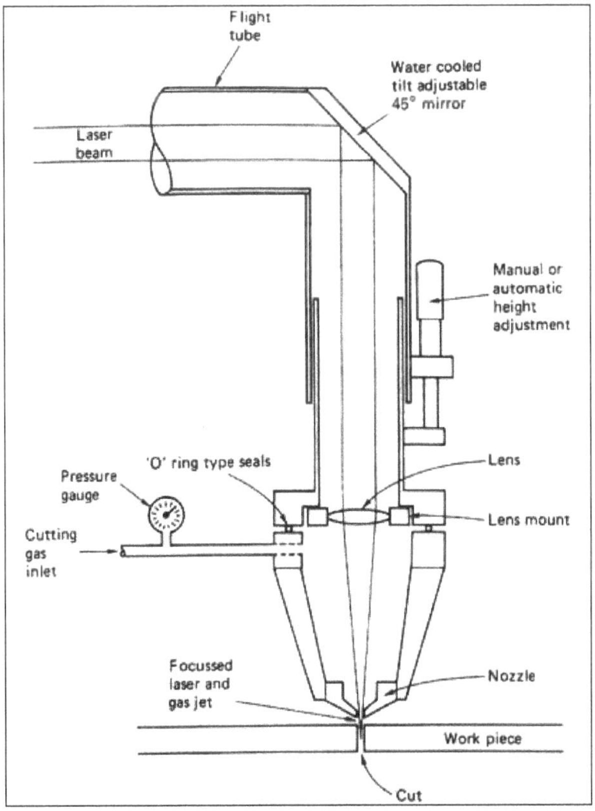

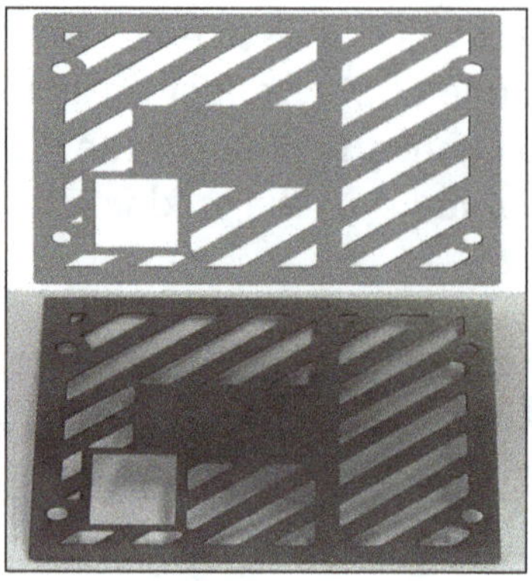

CAD (top) and stainless steel laser-cut part.

Process

Generation of the laser beam involves stimulating a lasing material by electrical discharges or lamps within a closed container. As the lasing material is stimulated, the beam is reflected internally by means of a partial mirror, until it achieves sufficient energy to escape as a stream of monochromatic coherent light. Mirrors or fiber optics are typically used to direct the coherent light to a lens, which focuses the light at the work zone. The narrowest part of the focused beam is generally less than 0.0125 inches (0.32 mm). in diameter. Depending upon material thickness, kerf widths as small as 0.004 inches (0.10 mm) are possible. In order to be able to start cutting from somewhere other than the edge, a pierce is done before every cut. Piercing usually involves a high-power pulsed laser beam which slowly makes a hole in the material, taking around 5–15 seconds for 0.5-inch-thick (13 mm) stainless steel.

Industrial laser cutting of steel with cutting instructions programmed through the CNC interface.

The parallel rays of coherent light from the laser source often fall in the range between 0.06–0.08 inches (1.5–2.0 mm) in diameter. This beam is normally focused and intensified by a lens or a mirror to a very small spot of about 0.001 inches (0.025 mm) to create a very intense laser beam. In order to achieve the smoothest possible finish during contour cutting, the direction of beam polarization must be rotated as it goes around the periphery of a contoured workpiece. For sheet metal cutting, the focal length is usually 1.5–3 inches (38–76 mm).

Advantages of laser cutting over mechanical cutting include easier workholding and reduced contamination of workpiece (since there is no cutting edge which can become contaminated by the material or contaminate the material). Precision may be better, since the laser beam does not wear during the process. There is also a reduced chance of warping the material that is being cut, as laser systems have a small heat-affected zone. Some materials are also very difficult or impossible to cut by more traditional means.

Laser cutting for metals has the advantages over plasma cutting of being more precise and using less energy when cutting sheet metal; however, most industrial lasers cannot cut through the greater metal thickness that plasma can. Newer laser machines operating at higher power (6000 watts, as contrasted with early laser cutting machines' 1500 watt ratings) are approaching plasma machines in their ability to cut through thick materials, but the capital cost of such machines is much higher than that of plasma cutting machines capable of cutting thick materials like steel plate.

Types

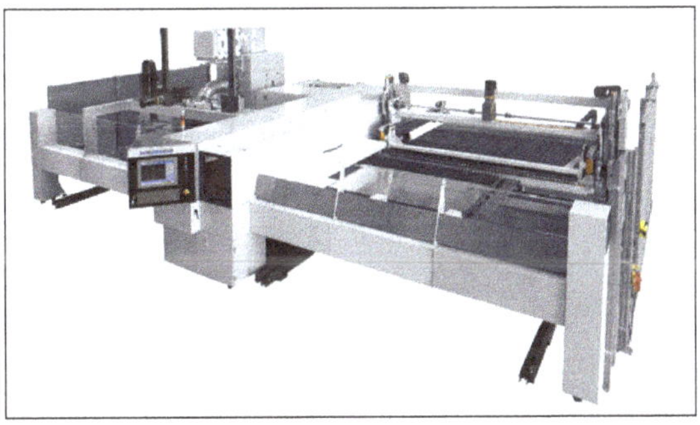

HACO fiber laser cutting machine with an integrated loading and unloading system.

4000 watt CO_2 laser cutter.

There are three main types of lasers used in laser cutting. The CO_2 laser is suited for cutting, boring, and engraving. The neodymium (Nd) and neodymium yttrium-aluminium-garnet (Nd:YAG) lasers are identical in style and differ only in application. Nd is used for boring and where high energy but low repetition are required. The Nd:YAG laser is used where very high power is needed and for boring and engraving. Both CO_2 and Nd/Nd:YAG lasers can be used for welding.

CO_2 lasers are commonly "pumped" by passing a current through the gas mix (DC-excited) or using radio frequency energy (RF-excited). The RF method is newer and has become more popular. Since DC designs require electrodes inside the cavity, they can encounter electrode erosion and plating of electrode material on glassware and optics. Since RF resonators have external electrodes they are not prone to those problems. CO_2 lasers are used for industrial cutting of many materials including titanium, stainless steel, mild steel, aluminium, plastic, wood, engineered wood, wax, fabrics, and paper. YAG lasers are primarily used for cutting and scribing metals and ceramics.

In addition to the power source, the type of gas flow can affect performance as well. Common variants of CO_2 lasers include fast axial flow, slow axial flow, transverse flow, and slab. In a fast axial flow resonator, the mixture of carbon dioxide, helium and nitrogen is circulated at high velocity by a turbine or blower. Transverse flow lasers circulate the gas mix at a lower velocity, requiring a simpler blower. Slab or diffusion cooled resonators have a static gas field that requires no pressurization or glassware, leading to savings on replacement turbines and glassware.

The laser generator and external optics (including the focus lens) require cooling. Depending on system size and configuration, waste heat may be transferred by a coolant or directly to air. Water is a commonly used coolant, usually circulated through a chiller or heat transfer system.

A *laser microjet* is a water-jet guided laser in which a pulsed laser beam is coupled into a low-pressure water jet. This is used to perform laser cutting functions while using the water jet to guide the laser beam, much like an optical fiber, through total internal reflection. The advantages of this are that the water also removes debris and cools the material. Additional advantages over traditional "dry" laser cutting are high dicing speeds, parallel kerf, and omnidirectional cutting.

Fiber lasers are a type of solid state laser that is rapidly growing within the metal cutting industry. Unlike CO_2, Fiber technology utilizes a solid gain medium, as opposed to a gas or liquid. The "seed laser" produces the laser beam and is then amplified within a glass fiber. With a wavelength of only 1.064 micrometers fiber lasers produce an extremely small spot size (up to 100 times smaller compared to the CO_2) making it ideal for cutting reflective metal material. This is one of the main advantages of Fiber compared to CO_2.

Methods

There are many different methods in cutting using lasers, with different types used to cut different material. Some of the methods are vaporization, melt and blow, melt blow and burn, thermal stress cracking, scribing, cold cutting and burning stabilized laser cutting.

Vaporization Cutting

In vaporization cutting the focused beam heats the surface of the material to boiling point and generates a keyhole. The keyhole leads to a sudden increase in absorptivity quickly deepening the hole. As the hole deepens and the material boils, vapor generated erodes the molten walls blowing ejecta out and further enlarging the hole. Non melting material such as wood, carbon and thermoset plastics are usually cut by this method.

Melt and Blow

Melt and blow or fusion cutting uses high-pressure gas to blow molten material from the cutting area, greatly decreasing the power requirement. First the material is heated to melting point then a gas jet blows the molten material out of the kerf avoiding the need to raise the temperature of the material any further. Materials cut with this process are usually metals.

Thermal Stress Cracking

Brittle materials are particularly sensitive to thermal fracture, a feature exploited in thermal stress cracking. A beam is focused on the surface causing localized heating and thermal expansion. This results in a crack that can then be guided by moving the beam. The crack can be moved in order of m/s. It is usually used in cutting of glass.

Stealth Dicing of Silicon Wafers

The separation of microelectronic chips as prepared in semiconductor device fabrication from silicon wafers may be performed by the so-called stealth dicing process, which operates with a pulsed Nd:YAG laser, the wavelength of which (1064 nm) is well adopted to the electronic band gap of silicon (1.11 eV or 1117 nm).

Reactive Cutting

Also called "burning stabilized laser gas cutting", "flame cutting". Reactive cutting is like oxygen torch cutting but with a laser beam as the ignition source. Mostly used for cutting carbon steel in thicknesses over 1 mm. This process can be used to cut very thick steel plates with relatively little laser power.

Tolerances and Surface Finish

Laser cutters have positioning accuracy of 10 micrometers and repeatability of 5 micrometers.

Standard roughness Rz increases with the sheet thickness, but decreases with laser power and cutting speed. When cutting low carbon steel with laser power of 800 W, standard roughness Rz is 10 μm for sheet thickness of 1 mm, 20 μm for 3 mm, and 25 μm for 6 mm.

$$Rz = \frac{12.528 \cdot S^{0.542}}{P^{0.528} \cdot V^{0.322}}$$

Where: S = steel sheet thickness in mm; P = laser power in kW (some new laser cutters have laser power of 4 kW); V = cutting speed in meters per minute.

This process is capable of holding quite close tolerances, often to within 0.001 inch (0.025 mm). Part geometry and the mechanical soundness of the machine have much to do with tolerance capabilities. The typical surface finish resulting from laser beam cutting may range from 125 to 250 micro-inches (0.003 mm to 0.006 mm).

Machine Configurations

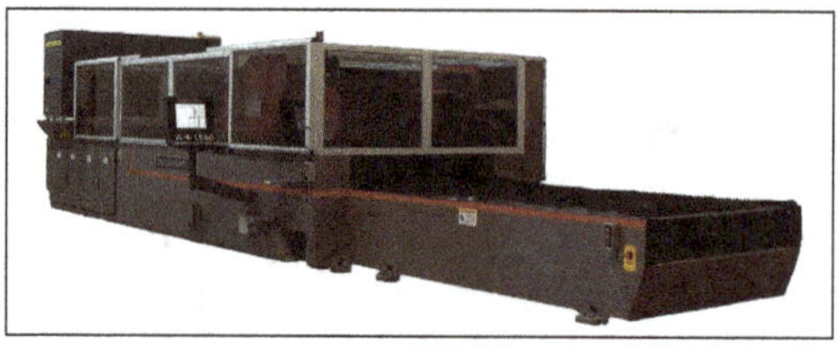

Dual-pallet flying optics laser.

There are generally three different configurations of industrial laser cutting machines: moving material, hybrid, and flying optics systems. These refer to the way that the laser beam is moved over the material to be cut or processed. For all of these, the axes of motion are typically designated X and Y axis. If the cutting head may be controlled, it is designated as the Z-axis.

Flying optics laser head.

Moving material lasers have a stationary cutting head and move the material under it. This method provides a constant distance from the laser generator to the workpiece and a single point from which to remove cutting effluent. It requires fewer optics, but requires moving the workpiece. This style machine tends to have the fewest beam delivery optics, but also tends to be the slowest.

Hybrid lasers provide a table which moves in one axis (usually the X-axis) and move the head along the shorter (Y) axis. This results in a more constant beam delivery path length than a flying optic machine and may permit a simpler beam delivery system. This can result in reduced power loss in the delivery system and more capacity per watt than flying optics machines.

Flying optics lasers feature a stationary table and a cutting head (with laser beam) that moves over the workpiece in both of the horizontal dimensions. Flying optics cutters keep the workpiece stationary during processing and often do not require material clamping. The moving mass is constant, so dynamics are not affected by varying size of the workpiece. Flying optics machines are the fastest type, which is advantageous when cutting thinner workpieces.

Flying optic machines must use some method to take into account the changing beam length from near field (close to resonator) cutting to far field (far away from resonator) cutting. Common methods for controlling this include collimation, adaptive optics or the use of a constant beam length axis.

Five and six-axis machines also permit cutting formed workpieces. In addition, there are various methods of orienting the laser beam to a shaped workpiece, maintaining a proper focus distance and nozzle standoff, etc.

Pulsing

Pulsed lasers which provide a high-power burst of energy for a short period are very effective in some laser cutting processes, particularly for piercing, or when very small holes or very low cutting speeds are required, since if a constant laser beam were used, the heat could reach the point of melting the whole piece being cut.

Most industrial lasers have the ability to pulse or cut CW (continuous wave) under NC (numerical control) program control.

Double pulse lasers use a series of pulse pairs to improve material removal rate and hole quality. Essentially, the first pulse removes material from the surface and the second prevents the ejecta from adhering to the side of the hole or cut.

Power Consumption

The main disadvantage of laser cutting is the high power consumption. Industrial laser efficiency may range from 5% to 45%. The power consumption and efficiency of any particular laser will vary depending on output power and operating parameters. This will depend on type of laser and how well the laser is matched to the work at hand. The amount of laser cutting power required, known as *heat input*, for a particular job depends on the material type, thickness, process (reactive/inert) used, and desired cutting rate.

Table: Amount of heat input required for various material at various thicknesses using a CO_2 laser watts.

Material	Material thickness				
	0.51 mm	1.0 mm	2.0 mm	3.2 mm	6.4 mm
Stainless steel	1000	1000	1000	1500	2500
Aluminium	1000	1000	1000	3800	10000
Mild steel	–	400	–	500	–
Titanium	250	210	210	–	–
Plywood	–	–	–	–	650
Boron/epoxy	–	–	–	3000	–

Production and Cutting Rates

The maximum cutting rate (production rate) is limited by a number of factors including laser power, material thickness, process type (reactive or inert), and material properties. Common industrial systems (≥1 kW) will cut carbon steel metal from 0.51 – 13 mm in thickness. For all intents and purposes, a laser can be up to thirty times faster than standard sawing.

Table: Cutting rates using a CO_2 laser cm/second.

Workpiece material	Material thickness					
	0.51 mm	1.0 mm	2.0 mm	3.2 mm	6.4 mm	13 mm
Stainless steel	42.3	23.28	13.76	7.83	3.4	0.76
Aluminium	33.87	14.82	6.35	4.23	1.69	1.27
Mild steel	–	8.89	7.83	6.35	4.23	2.1
Titanium	12.7	12.7	4.23	3.4	2.5	1.7
Plywood	–	–	–	–	7.62	1.9
Boron / epoxy	–	–	–	2.5	2.5	1.1

Laser Beam Welding

Laser Beam Welding is a fusion welding process in which two metal pieces are joined together by the use of laser. The laser beams are focused to the cavity between the two metal pieces to be joined. The laser beams have enough energy and when it strikes the metal pieces produces heat that melts the material from the two metal pieces and fills the cavity. After cooling a strong weld is formed between the two pieces.

Working Principle

It works on the principle that when electrons of an atom gets excited by absorbing some energy. And then after some time when it returns back to its ground state, it emits a photon of light. The concentration of this emitted photon increased by stimulated emission of radiation and we get a high energy concentrated laser beam. Light amplification by stimulated emission of radiation is called laser.

Main Parts

The main parts or equipment of laser beam welding are:

- Laser Machine: It is a machine that is used to produce laser for welding. The main components of laser machine are shown below.
- Power Source: A high voltage power source is applied across the laser machine to produce laser beam.
- CAM: It is a computer aided manufacturing in which the laser machine is integrated with the computers to perform welding process. All the controlling action during the welding process by laser is done by CAM. It speeds up the welding process to a greater extent.
- CAD: It is called as Computer aided Design. It is used to design the job for welding. Here computers are used to design the workpiece and how the welding is performed on it.
- Shielding Gas: A shielding gas may be used during the welding process in order to prevent the w/p from oxidation.

Types of Laser Used

- Gas lasers: It uses mixtures of gases as lasing medium to produce laser. Mixtures of gases such as nitrogen, helium and Co_2 are used as lasing medium.
- Solid-state laser: it uses several solid media such as synthetic ruby crystal (chromium in aluminum oxide), neodymium in glass (Nd:glass), and neodymium in yttrium aluminum garnet (Nd-YAG, most commonly used).
- Fiber laser: The lasing medium in this type of laser is optical fiber itself.

Characteristics of Laser Beam Welding

- The power density of laser beam welding is high. It is of the order 1 MW/cm^2. Because of this high energy density, it has small heat-affected zones. The rate of heating and cooling is high.
- The laser beams produced are coherent (having same phase) and monochromatic (i.e. having same wavelength).
- It is used to weld smaller sizes spot but the spot sizes can vary from. 2mm to 13 mm.
- The depth of penetration of the LBW depends upon the amount of power supply and location of the focal point. It is proportional the amount of power supply. When the focal point is kept slightly below the surface of the workpiece, the depth of penetration is maximized.

- Pulsed or continuous laser beams are used for welding. Thin materials are weld by using millisecond-pulses and continuous laser beams are used for deep welds.
- It is versatile process because it is capable of welding carbon steels, stainless steel, HSLA Steels, aluminum and titanium. Due to high cooling rate, the problem of cracking is there when welding high-carbon steels.
- It produces high quality weld.
- This welding process is most popular in automotive industry.

Working

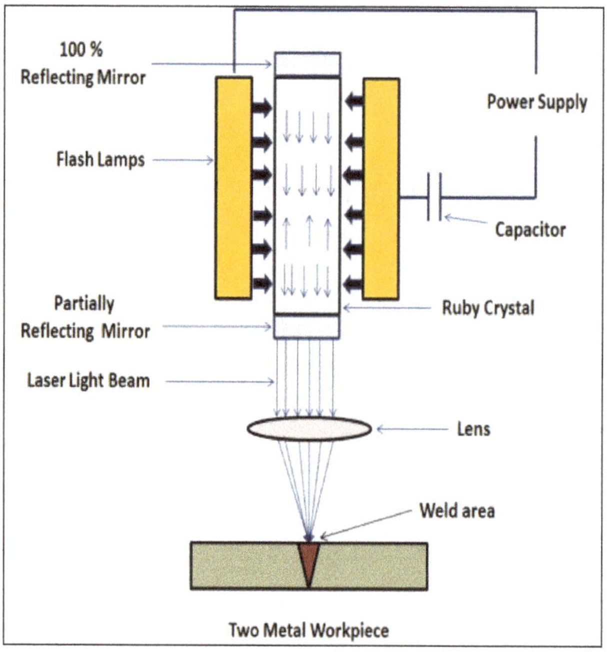

- First the setup of welding machine at the desired location (in between the two metal pieces to be joined) is done.
- After setup, a high voltage power supply is applied on the laser machine. This starts the flash lamps of the machine and it emits light photons. The energy of the light photon is absorbed by the atoms of ruby crystal and electrons get excited to their higher energy level. When they return back to their ground state (lower Energy state) they emit a photon of light. This light photon again stimulates the excited electrons of the atom and produces two photons. This process keeps continue and we get a concentrated laser beam.
- This high concentrated laser beam is focused to the desired location for the welding of the multiple pieces together. Lens are used to focus the laser to the

area where welding is needed. CAM is used to control the motion of the laser and workpiece table during the welding process.

- As the laser beam strikes the cavity between the two metal pieces to be joined, it melts the base metal from both the pieces and fuses them together. After solidification we get a strong weld.
- This is how a laser Beam Welding Works.

Advantages

- It produces high weld quality.
- LBW can be easily automated with robotic machinery for large volume production.
- No electrode is required.
- No tool wears because it is a non-contact process.
- The time taken for welding thick section is reduced.
- It is capable of welding in those areas which is not easily accessible.
- It has the ability to weld metals with dissimilar physical properties.
- It can be weld through air and no vacuum is required.
- X – Ray shielding is not required as it does not produce any X-Rays.
- It can be focused on small areas for welding. This is because of its narrower beam of high energy.
- Wide variety of materials can be welded by using laser beam welding.
- It produces weld of aspect ratio (i.e. depth to width ratio) of 10:1.

Disadvantages

- Initial cost is high. The equipment used in LBW has high cost.
- High maintenance cost.
- Due to rapid rate of cooling, cracks may be produced in some metals.
- High skilled labour is required to operate LBW.
- The welding thickness is limited to 19 mm.
- The energy conversion efficiency in LBW is very low. It is usually below 10 %.

Application

The laser beam welding is dominant in automotive industry. It is used in the area where large volume production is required.

Laser Pointers

A laser pointer is a small (usually battery-powered) laser device designed for pointing at objects by illuminating them with a collimated visible laser beam. Most laser pointers, particularly the cheap ones, contain a small GaInP/AlGaInP laser diode operating somewhere in the red spectral region, a collimating lens, a simple electronic diode driver, and a battery compartment for e.g. three coin cells. Some significantly more expensive pointers, as shown in figure, emit green or even blue or yellow light and normally contain a small diode-pumped solid-state laser with a nonlinear crystal for frequency doubling. Green laser pointers are usually based on a miniature Nd:YVO4 laser with a KTP crystal for intracavity frequency doubling. Here, Nd:YVO4 is beneficial for a low threshold pump power, and KTP works in a relatively wide temperature range, thus not requiring means for temperature stabilization.

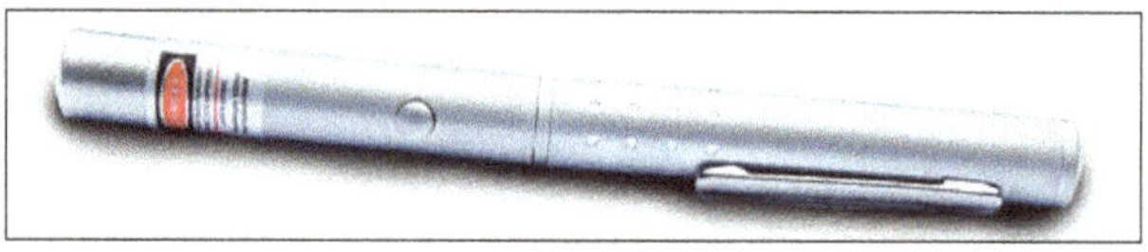

A green-emitting laser pointer, containing a tiny diode-pumped frequency-doubled solid-state laser. Red laser pointers are available in smaller sizes, because they do not need as large batteries.

Laser pointers should not be confused with lamps containing light-emitting diodes (LEDs), which emit a much more diffuse beam (with much lower spatial coherence, similar to that of an incandescent lamp) and can also emit light with different colors, or white light.

Applications

A typical use of a hand-held laser pointer is to point at some screen or chart during a presentation, e.g. a conference talk. This is convenient because it can be done from a large distance and requires only a small hand-held device. However, the visibility of the generated spot on the screen is often poor (particularly for red laser pointers with relatively long emission wavelength), and a fast-moving light spot can have a somewhat nervous appearance. Therefore, some people prefer an old-fashioned telescopic pointing device for presentations. Laser pointers can be useful for, e.g., aligning some machinery, or for certain optical distance measurements.

Brightness and Color

The apparent brightness of the illuminated spot depends strongly on the wavelength of the emitted light. Most devices operate in the red spectral region, where the sensitivity of the eye rapidly decreases with increasing wavelength. Devices with 650-nm output appear about twice as bright as those emitting the same power at 670 nm, and 635-nm devices still about two times brighter. However, the shorter-wavelength laser pointers are typically more expensive. This is particularly true for green lasers, which are significantly brighter than their red counterparts, but are still expensive. They involve a diode-pumped solid-state laser and a frequency doubler. Due to the typically poor conversion efficiency of the frequency doubler at low power levels, hundreds of milliwatts of infrared (typically 1064-nm) light are required for generating a few milliwatts in the green, and the batteries will accordingly not last very long, unless they are comparatively heavy.

Range of a Laser Pointer

Lay persons often ask what is the range of a laser pointer is, and responding to this interest some producers specify some more or less questionable numbers. If the question is meant to be how far the light of a laser pointer can propagate, the correct answer is that there is no limit, provided that the light is not absorbed or scattered away in the atmosphere. However, the beam area will eventually become larger due to the beam divergence, so that the intensity e.g. on a screen will be reduced even if the overall power remains constant. Accordingly, an airplane pilot looking down into such a beam from an altitude of 10 km will not be disturbed by the remaining small intensity.

The range of a laser pointer may also be understood as the maximum distance from which the spot on the screen can be seen. That kind of range is normally not limited by the beam divergence but by the overall optical power (apart from the wavelength and level of ambient light), since the issue is not the comparatively minor divergence on the way from the laser pointer to the screen, but rather the huge divergence of the scattered light on the way back. Therefore, someone standing next to the illuminated screen would easily see the spot when it is already hardly perceivable from the position of the laser pointer.

Safety Hazards

There have been extensive debates on laser safety issues associated with laser pointers. Typical output powers are a few milliwatts – normally below 5 mW in order to comply with safety class 3R, and sometimes below 1 mW for class 2. Direct staring into a 1-mW beam can be irritating for the eye: it can cause temporary flash blindness. However, nobody would normally do that long enough to cause serious eye damage. Nevertheless, great care should be taken, e.g. when children are playing with laser pointers, if laser pointers are at all considered to be suitable as toys. Significant hazards could arise e.g. if somebody walking down stairs or a car driver is irritated by a laser beam.

There are some reports saying that cheap green laser pointers are sold which do not have a filter to eliminate the infrared light, and therefore can emit hundreds of milliwatts in the infrared spectral region. This is obviously a terrible safety hazard; an eye directly hit by such a beam could be destroyed within a fraction of a second.

Use for Alignment

In principle, laser pointers can be used for alignment purposes. However, there are special alignment lasers which may be better suited for such applications. For example, they have a housing which can be more conveniently mounted, and the beam direction may be more stable.

Laser Drilling

Laser drilling is the process of creating thru-holes, referred to as "popped" holes or "percussion drilled" holes, by repeatedly pulsing focused laser energy on a material. The diameter of these holes can be as small as 0.002". If larger holes are required, the laser is moved around the circumference of the "popped" hole until the desired diameter is created; this technique is called "trepanning.

Applications

Laser drilling is one of the few techniques for producing high-aspect-ratio holes—holes with a depth-to-diameter ratio much greater than 10:1.

Laser-drilled high-aspect-ratio holes are used in many applications, including the oil gallery of some engine blocks, aerospace turbine-engine cooling holes, laser fusion components, and printed circuit board micro-vias.

Manufacturers of turbine engines for aircraft propulsion and for power generation have benefited from the productivity of lasers for drilling small (0.3–1 mm diameter typical) cylindrical holes at 15–90° to the surface in cast, sheet metal and machined components. Their ability to drill holes at shallow angles to the surface at rates of between 0.3 and 3 holes per second has enabled new designs incorporating film-cooling holes for improved fuel efficiency, reduced noise, and lower NOx and CO emissions.

Incremental improvements in laser process and control technologies have led to substantial increases in the number of cooling holes used in turbine engines. Fundamental to these improvements and increased use of laser drilled holes is an understanding of the relationship between process parameters and hole quality and drilling speed.

Theory

Following is a summary of technical insights about the laser drilling process and the relationship between process parameters and hole quality and drilling speed.

Physical Phenomena

Laser drilling of cylindrical holes generally occurs through melting and vaporization (also referred to as "ablation") of the workpiece material through absorption of energy from a focused laser beam. The energy required to remove material by melting is about

25% of that needed to vaporize the same volume, so a process that removes material by melting is often favored.

Whether melting or vaporization is more dominant in a laser drilling process depends on many factors, with laser pulse duration and energy playing an important role. Generally speaking, ablation dominates when a Q-switched Nd:YAG laser is used. On the other hand, melt expulsion, the means by which a hole is created through melting the material, dominates when a flashtube pumped Nd:YAG laser is used. A Q-switched Nd:YAG laser normally has pulse duration in the order of nanoseconds, peak power on the order of ten to hundreds of MW/cm^2, and a material removal rate of a few micrometres per pulse. A flash lamp pumped Nd:YAG laser normally has a pulse duration on the order of hundreds of microseconds to a millisecond, peak power in the order of sub MW/cm^2, and material removal rate of ten to hundreds of micrometers per pulse. For machining processes by each laser, ablation and melt expulsion typically coexist.

Melt expulsion arises as a result of the rapid build-up of gas pressure (recoil force) within a cavity created by evaporation. For melt expulsion to occur, a molten layer must form and the pressure gradients acting on the surface due to vaporization must be sufficiently large to overcome surface tension forces and expel the molten material from the hole.

The "best of both worlds" is a single system capable of both "fine" and "coarse" melt expulsion. "Fine" melt expulsion produces features with excellent wall definition and small heat-affected zone while "coarse" melt expulsion, such as used in percussion drilling and trepanning, removes material quickly.

The recoil force is a strong function of the peak temperature. The value of T_{cr} for which the recoil and surface tension forces are equal is the critical temperature for liquid expulsion. For instance, liquid expulsion from titanium can take place when the temperature at the center of the hole exceeds 3780 K.

In early work, the proportion of material removed by melt expulsion was found to increase as intensity increased. More recent work shows that the fraction of the material removed by melt expulsion, referred to as melt ejection fraction (MEF), drops when laser energy further increases. The initial increase in melt expulsion on raising the beam power has been tentatively attributed to an increase in the pressure and pressure gradient generated within the hole by vaporization.

A better finish can be achieved if the melt is ejected in fine droplets. Generally speaking, droplet size decreases with increasing pulse intensity. This is due to the increased vaporization rate and thus a thinner molten layer. For the longer pulse duration, the greater total energy input helps form a thicker molten layer and results in the expulsion of correspondingly larger droplets.

Previous Models

Chan and Mazumder developed a 1-D steady state model to incorporate liquid expulsion consideration but the 1-D assumption is not suited for high aspect ratio hole drilling and the drilling process is transient. Kar and Mazumder extended the model to 2-D, but melt expulsion was not explicitly considered. A more rigorous treatment of melt expulsion has been presented by Ganesh, et al., which is a 2-D transient generalized model to incorporate solid, fluid, temperature, and pressure during laser drilling, but it is computationally demanding. Yao, et al. developed a 2-D transient model, in which a Knudsen layer is considered at the melt-vapor front, and the model is suited for shorter pulse and high peak power laser ablation.

Laser Energy Absorption and Melt-vapor Front

At the melt-vapor front, the Stefan boundary condition is normally applied to describe the laser energy absorption.

$$I_{abs} + k\left(\frac{\partial T}{\partial z} + r\frac{\partial T}{\partial r}\right) + \rho_l V_i L_v - \rho_v V_v (c_p T_i + E_v) = 0$$

where $I_{abs} = I(t)^{-\beta z}$ is the absorbed laser intensity, β is the laser absorption coefficient depending on laser wavelength and target material, and *I(t)* describes temporal input laser intensity including pulse width, repetition rate, and pulse temporal shape. *k* is the heat conductivity, *T* is the temperature, *z* and *r* are distances along axial and radial directions, *p* is density, *v* the velocity, L_v the latent heat of vaporization. The subscripts *l*, *v* and *i* denote liquid phase, vapor phase and vapor-liquid interface, respectively.

If the laser intensity is high and pulse duration is short, the so-called Knudsen layer is assumed to exist at the melt-vapor front where the state variables undergo discontinuous changes across the layer. By considering the discontinuity across the Knudsen layer, Yao, et al. simulated the surface recess velocity V_v distribution, along the radial direction at different times, which indicates the material ablation rate is changing significantly across the Knudsen layer.

Melt Expulsion

After obtaining the vapor pressure p_v, the melt layer flow and melt expulsion can be modeled using hydrodynamic equations. Melt expulsion occurs when the vapor pressure is applied on the liquid free surface which in turn pushes the melt away in the radial direction. In order to achieve fine melt expulsion, the melt flow pattern needs to be predicted very precisely, especially the melt flow velocity at the hole's edge. Thus, a 2-D axisymmetric transient model is used and accordingly the momentum and continuity equations used.

Ganesh's model for melt ejection is comprehensive and can be used for different stages of the hole drilling process. However, the calculation is very time consuming and Solana, et al., presented a simplified time dependent model that assumes that the melt expulsion velocity is only along the hole wall, and can give results with a minimum computational effort.

The liquid will move upwards with velocity u as a consequence of the pressure gradient along the vertical walls, which is given in turn by the difference between the ablation pressure and the surface tension divided by the penetration depth x.

Assuming that the drilling front is moving at a constant velocity, the following linear equation of liquid motion on the vertical wall is a good approximation to model the melt expulsion after the initial stage of drilling.

$$\rho\frac{\partial u(r,t)}{\partial t} = P(t) + \mu\frac{\partial^2 u(r,t)}{\partial r^2}$$

Where, p is the melt density, μ is the viscosity of the liquid, $P(t)=(\Delta P(t)/x(t))$ is the pressure gradient along the liquid layer, $\Delta P(t)$ is the difference between the vapor pressure P_v and the surface tension $\frac{2\sigma}{\bar{\delta}}$.

Pulse Shape Effect

Roos showed that a 200 μs train consisting of 0.5 μs pulses produced superior results for drilling metals than a 200 μs flat shaped pulse. Anisimov, et al. discovered that process efficiency improved by accelerating the melt during the pulse.

Grad and Mozina further demonstrated the effect of pulse shapes. A 12 ns spike was added at the beginning, middle, and the end of a 5 ms pulse. When the 12 ns spike was added to the beginning of the long laser pulse, where no melt had been produced, no significant effect on removal was observed. On the other hand, when the spike was added at the middle and the end of the long pulse, the improvement of the drilling efficiency was 80 and 90%, respectively. The effect of inter-pulse shaping has also been investigated. Low and Li showed that a pulse train of linearly increasing magnitude had a significant effect on expulsion processes. Forsman, et al. demonstrated that a double pulse stream produced increased drilling and cutting rates with significantly cleaner holes.

Laser Engraving

Laser engraving, which is a subset of laser marking, is the practice of using lasers to engrave an object. Laser marking, on the other hand, is a broader category of methods to leave marks on an object, which also includes color change due to chemical/molecular alteration, charring, foaming, melting, ablation, and more. The technique does not involve the use of inks, nor does it involve tool bits which contact the engraving surface and wear out, giving it an advantage over alternative engraving or marking technologies where inks or bit heads have to be replaced regularly.

The impact of laser marking has been more pronounced for specially designed "laser-able" materials and also for some paints. These include laser-sensitive polymers and novel metal alloys.

Laser marked electronic part.

The term laser marking is also used as a generic term covering a broad spectrum of surfacing techniques including printing, hot-branding and laser bonding. The machines for laser engraving and laser marking are the same, so that the two terms are sometimes confused by those without knowledge or experience in the practice.

Laser Engraving Machines

A laser engraving machine can be thought of as three main parts: a laser, a controller, and a surface. The laser is like a pencil - the beam emitted from it allows the controller to trace patterns onto the surface. The controller direction, intensity, speed of movement, and spread of the laser beam aimed at the surface. The surface is picked to match what the laser can act on.

A laser engraving machine.

There are three main genres of engraving machines: The most common is the X-Y table where, usually, the workpiece (surface) is stationary and the laser optics move around in X and Y directions, directing the laser beam to draw vectors. Sometimes the laser is stationary and the workpiece moves. Sometimes the workpiece moves in the Y axis and the laser in the X axis. A second genre is for cylindrical workpieces (or flat workpieces mounted around a cylinder) where the laser effectively traverses a fine helix and on/off laser pulsing produces the desired image on a raster basis. In the third method, both the laser and workpiece are stationary and galvo mirrors move the laser beam over the workpiece surface. Laser engravers using this technology can work in either raster or vector mode.

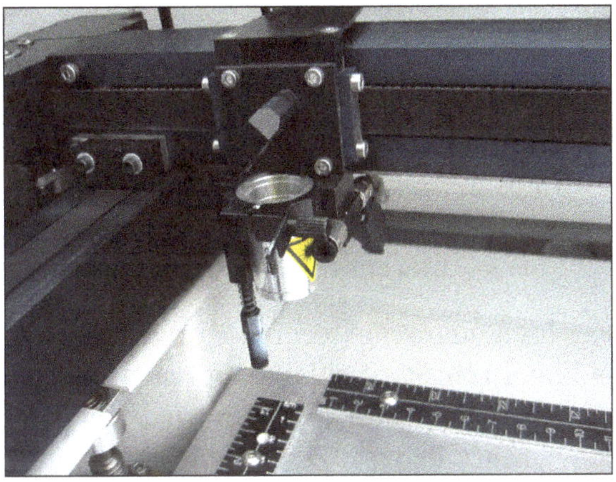

A laser engraver.

The point where the laser (the terms "laser" and "laser beam" may be used interchangeably) touches the surface should be on the focal plane of the laser's optical system and is usually synonymous with its focal point. This point is typically small, perhaps less than a fraction of a millimetre (depending on the optical wavelength). Only the area inside this focal point is significantly affected when the laser beam passes over the surface. The energy delivered by the laser changes the surface of the material at the focal point. It may heat up the surface and subsequently vaporize the material, or perhaps the material may fracture (known as "glassing" or "glassing up") and flake off the surface. Cutting through the paint of a metal part is generally how material is laser engraved.

If the surface material is vaporized during laser engraving, ventilation through the use of blowers or a vacuum pump are almost always required to remove the noxious fumes and smoke arising from this process, and for removal of debris on the surface to allow the laser to continue engraving.

A laser can remove material very efficiently because the laser beam can be designed to deliver energy to the surface in a manner which converts a high percentage of the light energy into heat. The beam is highly focused and collimated - in most non-reflective materials like wood, plastics and enamel surfaces, the conversion of light energy to heat is more than {x%} efficient. However, because of this efficiency, the equipment used in laser engraving may heat up rather quickly. Elaborate cooling systems are required for the laser. Alternatively, the laser beam may be pulsed to decrease the amount of excessive heating.

Different patterns can be engraved by programming the controller to traverse a particular path for the laser beam over time. The *trace* of the laser beam is carefully regulated to achieve a consistent removal depth of material. For example, criss-crossed paths are avoided to ensure that each etched surface is exposed to the laser only once, so the same amount of material is removed. The speed at which the beam moves across the material is also considered in creating engraving patterns. Changing the intensity and spread of the beam allows more flexibility in the design. For example, by changing the proportion of time (known as "duty-cycle") the laser is turned on during each pulse, the power delivered to the engraving surface can be controlled appropriately for the material.

Since the position of the laser is known exactly by the controller, it is not necessary to add barriers to the surface to prevent the laser from deviating from the prescribed engraving pattern. As a result, no resistive mask is needed in laser engraving. This is primarily why this technique is different from older engraving methods.

A good example of where laser engraving technology has been adopted into the industry norm is the production line. In this particular setup, the laser beam is directed towards a rotating or vibrating mirror. The mirror moves in a manner which may trace out numbers and letters onto the surface being marked. This is particularly useful for printing dates, expiry codes, and lot numbering of products travelling along a production line. Laser marking allows materials made of plastic and glass to be marked "on the move". The location where the marking takes place is called a "marking laser station", an entity often found in packaging and bottling plants. Older, slower technologies such as hot stamping and pad printing have largely been phased out and replaced with laser engraving.

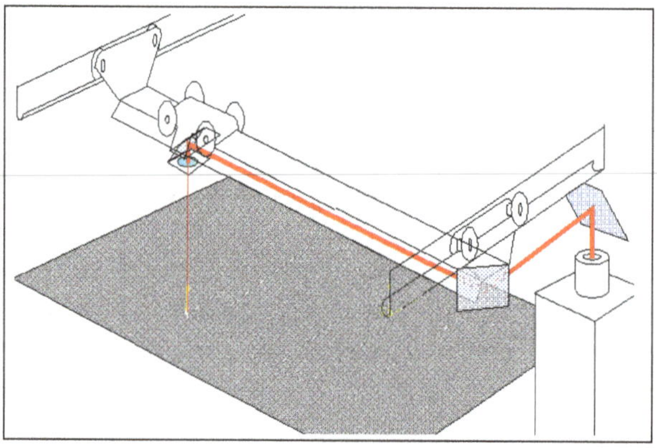

Mirrors on both X and Y carriages allow exact positioning.

For more precise and visually decorative engravings, a laser table is used. A laser table (or "X-Y table") is a sophisticated setup of equipment used to guide the laser beam more precisely. The laser is usually fixed permanently to the side of the table and emits light towards a pair of movable mirrors so that every point of the table surface can be swept by the laser. At the point of engraving, the laser beam is focused through a lens at the engraving surface, allowing very precise and intricate patterns to be traced out.

A typical setup of a laser table involves the fixed laser emitting light parallel to one axis of the table aimed at a mirror mounted on the end of an adjustable rail. The beam reflects off the mirror angled at 45 degrees so that the laser travels a path exactly along the length of the rail. This beam is then reflected by another mirror mounted to a movable trolley which directs the beam perpendicular to the original axis. In this scheme, two degrees of freedom (one vertical, and one horizontal) for etching can be represented.

In other laser engraving devices such as *flat table* or *drum engraving*, the laser beam is controlled to direct most of its energy a fixed penetration depth into the material to be engraved. In this manner, only a particular depth of material is removed when the engraving takes place. A simple machined stick or angle-iron can be used as a tool to help trained technologists adjust the engraver to achieve the required focusing. This setup is preferred for surfaces which do not vary in height appreciably.

For surfaces that vary in height, more elaborate focusing mechanisms have been developed. Some are known as *dynamic auto focus systems*. They adjust the lasing parameters in real time to adapt to the changes to the material as it is being etched. Typically, the height and depth of the surface are monitored with devices tracking changes to ultrasound, infrared, or visible light aimed at the engraving surface. These devices, known as *pilot beams* or *pilot lasers* (if a laser is used) help guide the adjustments made to the lens of the laser in determining the optimal spot to focus on the surface and remove material effectively.

"X-Y" laser engraving machines may operate in vector and raster mode. Vector engraving follows the line and curve of the pattern to be engraved, much like a pen-based plotter draws by constructing line segments from a description of the outlines of a pattern. Much early engraving of signs and plaques (laser or otherwise) used pre-stored font outlines so that letters, numbers or even logos could be scaled to size and reproduced with exactly defined strokes. Unfortunately, "fill" areas were problematic, as cross-hatching patterns and dot-fills sometimes exhibited moiré effects or uber-patterns caused by the imprecise calculation of dot spacings. Moreover, rotations of a font or dynamic scaling often were beyond the capabilities of the font-rendering device. The introduction of the PostScript page-description language now allows much greater flexibility—now virtually anything that can be described in vectors by PostScript-enabled software like CorelDRAW or Adobe Illustrator can be outlined, filled with suitable patterns, and laser-engraved.

Raster engraving traces the laser across the surface in a back-and-forth slowly advancing linear pattern that will remind one of the printhead on an inkjet or similar printer. The pattern is usually optimized by the controller/computer so that areas to either side of the pattern which aren't to be engraved are ignored and the trace across the material is thus shortened for better efficiency. The amount of advance of each line is normally less than the actual dot-size of the laser; the engraved lines overlap just slightly to create a continuity of engravure. As is true of all rasterized devices, curves and diagonals can sometimes suffer if the length or position of the raster lines varies even slightly in relation to the adjacent raster scan; therefore exact positioning and repeatability are critically important to the design of the machine. The advantage of rasterizing is the near effortless "fill" it produces. Most images to be engraved are bold letters or have large continuously engraved areas, and these are well-rasterized. Photos are rasterized (as in printing), with dots larger than that of the laser's spot, and these also are best engraved as a raster image. Almost any page-layout software can be used to feed a raster driver for an X-Y or drum laser engraver. While traditional sign and plaque engraving tended to favour the solid strokes of vectors out of necessity, modern shops tend to run their laser engravers mostly in raster mode, reserving vector for a traditional outline "look" or for speedily marking outlines or "hatches" where a plate is to be cut.

Materials that can be Engraved

Natural Materials

The marking of organic materials like wood is based on material carbonisation which produces darkening of the surface and marks with high contrast. Directly "burning" images on wood were some of the first uses of engraving lasers. The laser power required here is often less than 10 watts depending on the laser being used as most are different. Hardwoods like walnut, mahogany and maple produce good results. Softwoods can be judiciously engraved but tend to vaporize at less-consistent depths. Marking softwood requires the lowest power levels and enables the fastest cut speeds, while active cooling (e.g. a fan with sufficient airflow) inhibits ignition. Hard papers and fiberboard work well; linty papers and newsprint are like softwoods. Fur is not engraveable; finished leathers though can be laser-engraved with a look very similar to hot-branding. Certain latex rubber compounds can be laser engraved; for example these can be used to fabricate inking-stamps.

Paper masking tape is sometimes used as a pre-engraving overcoat on finished and resiny woods so that cleanup is a matter of picking the tape off and out of the unengraved areas, which is easier than removing the sticky and smoky surround "halos" (and requires no varnish-removing chemicals).

Plastics

Each plastic has specific material properties, especially the light absorption spectrum. The laser irradiation can generate direct chemical modifications, melting or evaporation of the material. Plastics are rarely seen in their pure state because several additives are used such as colorants, ultraviolet retardants, release agents, etc. These additives impact the result of laser marking.

Standard cast acrylic plastic, acrylic plastic sheet, and other cast resins generally laser very well. A commonly engraved award is a cast acrylic shape designed to be lasered from the back side. Styrene (as in compact disc cases) and many of the thermoforming plastics will tend to melt around the edge of the engraving spot. The result is usually "soft" and has no "etch" contrast. The surface may actually deform or "ripple" at the lip areas. In some applications this is acceptable; for example date markings on 2-litre soda bottles do not need to be sharp.

For signage and face plates, etc., special laser-marked plastics were developed. These incorporate silicate or other materials which conduct excess heat away from the material before it can deform.

Other plastics may be successfully engraved, but orderly experimentation on a sample piece is recommended. Bakelite is said to be easily laser-engraved; some hard engineering plastics work well. Expanded plastics, foams and vinyls, however, are generally candidates for routing rather than laser engraving. Plastics with a chlorine content (such as vinyl, PVC) produce corrosive chlorine gas when lasered, which combines with Hydrogen in the air to produce vaporized hydrochloric acid which can damage a laser engraving system. Urethane and silicone plastics usually don't work well—unless it is a formulation filled with cellulose, stone or some other stable insulator material.

Many light switchplates from companies such as Leviton or Lutron can be laser engraved. Again, experimentation may be necessary to develop the correct laser settings to result in engraving the surface rather than melting it. Often the laser engraving is followed by backfilling with paint on the engraved surface to produce more contrast between the engraved surface and the surrounding surface. Kevlar can be laser-engraved and laser-cut. However, Kevlar does give off extremely hazardous fumes (cyanide gas) when it is vaporized.

Metals

Metals are heat resistant materials, marking metals requires high-density laser irradiation. Basically, the average laser power leads to melting and the peak power causes evaporation of the material.

Laser on Stainless Steel (SS316L).

The best traditional engraving materials started out being the worst laser-engravable materials. This problem has now been solved using lasers at shorter wavelengths than the traditional 10,640 nm wavelength CO_2 laser. Using Yb: Fiber Lasers, Nd:YVO_4 or Nd:YAG lasers at 1,064 nm wavelength, or its harmonics at 532 and 355 nm, metals can now easily be engraved using commercial systems.

Coated Metals

The same conduction that works against the spot vaporization of metal is an asset if the objective is to vaporize some other coating away from the metal. Laser engraving metal plates are manufactured with a finely polished metal, coated with an enamel paint made to be "burned off". At levels of 10-30 watts, excellent engravings are made as the enamel is removed quite cleanly. Much laser engraving is sold as exposed brass or silver-coated steel lettering on a black or dark-enamelled background. A wide variety of finishes are now available, including screen-printed marble effects on the enamel. Anodized aluminum is commonly engraved or etched with CO_2 laser machines. With power less than 40W this metal can easily be engraved with clean, impressive detail. The laser bleaches the color exposing the white or silver aluminum substrate. Although it comes in various colors, laser engraving black anodized aluminum provides the best contrast of all colors. Unlike most materials engraving anodize aluminum does not leave any smoke or residue.Spray coatings can be obtained for the specific use of laser engraving metals, these sprays apply a coating that is visible to the laser light which fuses the coating to the substrate where the laser passed over. Typically, these sprays can also be used to engrave other optically invisible or reflective substances such as glass and are available in a variety of colours. Besides spray coatings, some laser-markable metals come pre-coated for imaging. Products such as this transform the surface of the metal to a different color (often black, brown or grey).

Stone and Glass

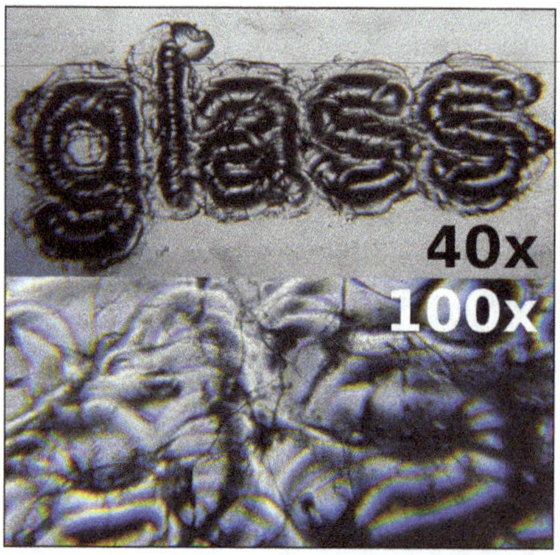

Laser engraved glass microscope slide with the word "glass" engraved in 3pt font. Magnified to 40x and 100x.

Stone and glass do not turn gaseous very easily. As expected, this makes them generally a better candidate for other means of engraving, most notably sandblasting or cutting using diamonds and water. But when a laser hits glass or stone, something else interesting happens: it fractures. Pores in the surface expose natural grains and crystalline "stubs" which, when heated very quickly, can separate a microscopic sized "chip" from the surface because the hot piece is expanding relative to its surroundings. So lasers are indeed used to engrave on glass, and if the power, speed and focus are just right, excellent results can be achieved. One should avoid large "fill" areas in glass engraving because the results across an expanse tend to be uneven; the glass ablation simply cannot be depended on for visual consistency, which may be a disadvantage or an advantage depending on the circumstances and the desired effect.

Jewelry

The demand for personalized jewelry has made jewelers more aware of the benefits of the laser engraving process. Jewellers found that by using a laser, they could tackle an engraving task with greater precision. In fact, jewelers discovered that laser engraving allowed for more precision than other types of engraving. At the same time, jewellers discovered that laser applied engravings had a number of other desirable features.

At one time jewellers who attempted to do laser engraving did need to use large pieces of equipment. Now the devices that perform laser engraving come in units. Some entrepreneurs have placed such units in mall kiosks. That has made laser engraving jewellery much more accessible. The makers of machines for laser engraving jewellers have developed some very specialized equipment. They have designed machines that can engrave the inside of a ring. They have also created machines that have the ability to engrave the back of a watch.

A laser can cut into both flat and curved surfaces such as the surfaces on jewellery. That points out the reason why jewellers have welcomed all the adaptations for the creation of laser engraved jewellery.

Fine Art

Laser engraving can also be used to create works of fine art. Generally, this involves engraving into planar surfaces, to reveal lower levels of the surface or to create grooves and striations which can be filled with inks, glazes, or other materials. Some laser engravers have rotary attachments which can engrave around an object.

Laser Etched Mirrors

Similarly, as with regular etched mirrors, the initial focus of Laser Engraving machines was to laser onto the glass surface of the mirror. When power, focus and speed are optimized, similar results to sandblasting or chemical etching can be achieved.

A new concept of Mirror Engraving is a laser engraved mirror. Here the laser pulsates through the reflective silver layer at the rear of the mirror. As a result the glass side of a Laser Engraved mirror remains intact, therefore maintaining the full reflective qualities of the original mirror.

Photograph of a laser engraved mirror. Mirror size 20x30 cm rectangle. Etched with photo and text engraving.

Finally, after this intricate laser process, the rear of the mirror needs to be "filled" with a new coating to replete the lasered detail in the mirror. When a photograph or text is laser engraved, a rear coating of solid black will give the best effect and will result in a defined black & white image. Alternatively colour coatings can be used giving a further dimension to the engraving.

Industrial Applications

Direct Laser Engraving of Flexographic Plates and Cylinders

Direct laser engraving of flexographic printing cylinders and plates has been an established process since the 1970s. This first began with the use of a carbon dioxide laser used to selectively ablate or evaporate a variety of rubber plate and sleeve materials to produce a print ready surface without the use of photography or chemicals. With this process there is no integral ablation mask as with direct photopolymer laser imaging. Instead a high-power carbon dioxide laser head burns away, or ablates, unwanted material. The aim is to form sharp, relief images with steep first relief and contoured shoulder supported edges to give a high standard of process color reproduction. A short water wash and dry cycle follows, which is a lot less involved than in the post-processing stages for direct laser imaging or conventional flexo platemaking using photopolymer plates. After engraving, the photopolymer is exposed through the imaged black layer and washed out in the traditional photopolymer process requiring photography and chemicals.

Before the year 2000 lasers only produced lower quality in rubber-like materials. In these rubber-like materials, which had a rough structure, higher quality was impossible. Since the year 2000 fiber lasers have been introduced to give a much increased engraving quality direct into black polymeric materials. Also at the Drupa 2004 the direct engraving of polymer plates was introduced. This had also an effect on the rubber-developers who, in order to stay competitive, developed new high quality rubber-like materials. The development of suitable polymeric compounds has also allowed the engraving quality achievable with the fibre lasers to be realised in print. Since then direct laser engraving of flexo-printingforms is seen by many as the modern way to make printing-forms for it is the first real digital way.

As a competitive process, more recently laser system have been introduced to selectively engrave the thin opaque black layer of a specially produced photopolymer plate or sleeve.

Direct Photopolymer Laser Imaging

Closely related is the direct imaging of a digital flexo plates or sleeves 'in-the-round' on a fast-rotating drum, or cylinder. This is carried out on a platesetter integrated within a digital prepress workflow, that also supports digital proofing. Again, this is a filmless process, which removes one of the variables in obtaining the fine and sharp dots for screened effects, including process color printing.

With this process the electronically generated image is scanned at speed to a photopolymer plate material that carries a thin black mask layer on the surface. The infrared laser-imaging head, which runs parallel to the drum axis, ablates the integral mask to reveal the uncured polymer underneath. A main ultraviolet exposure follows to form the image through the mask. The remaining black layer absorbs the ultraviolet radiation, which polymerizes the underlying photopolymer where the black layer has been removed. The exposed digital plate still needs to be processed like a conventional flexo plate. That is, using solvent-based washout with the necessary waste recovery techniques, although some water-washable digital plates are in development. This technology has been used since 1995 and is only now becoming more widely used around the world as more affordable equipment becomes available. Trade sources say there are around 650 digital platesetters installed in label, packaging and trade platemaking houses.

Laser Engraving of Anilox Rolls

Prior to 1980 anilox rolls were produced by a variety of mechanical processes. These metal anilox rolls were sometimes sprayed with ceramic to prolong their life in the flexographic printing press. During the 1980s laser engraving systems were produced which used a carbon dioxide laser to engrave the required cell pattern directly into the polished ceramic surface. Since then Q-switched YAG lasers were used for a period as they provided a more focusable laser beam as well as increased pulsing frequencies capable of engraving the finer cell configuration demanded by the ever-evolving flexographic printing process. Since approximately the year 2000 the direct anilox laser engraving process has been dominated by the use of fibre lasers which provide the high powers of the carbon dioxide lasers together with the finely focusable beam of the YAG lasers. Optical systems providing the rapid switching of multiple beams have allowed the fibre laser system to be dominant in this market. This technology has become known as Multi-Beam-Anilox or MBA.

Sub-surface Laser Engraving (SSLE)

Sub-surface laser engraving is the process of engraving an image in a transparent solid material by focusing a laser below the surface to create small fractures. Such engraved materials are of high-grade optical quality (suitable for lenses, with low dispersion) to minimize distortion of the beam. BK7 glass is a common material for this application. Plastics are also used, but with far less desirable results when compared to the engraving done in optical crystal.

Since its commercial application in the late 1990s, SSLE has become more cost effective with a number of different sized machines ranging from small (~US$35,000–60,000) to large production scale tables (>US$250,000). Although these machines are becoming more available, it is estimated that only a few hundred are in operation worldwide. Many machines require very expensive cooling, maintenance and calibration for proper use. The more popular SSLE engraving machines use the Diode Pumped Solid State or DPSS laser process. The laser diode, the primary component which excites a pulsed solid state laser, can easily cost one third of the machine itself and functions for a limited number of hours, although a good quality diode can last thousands of hours.

Since 2009, use of SSLE has become more cost effective to produce 3D images in souvenir 'crystal' or promotional items with only a few designers concentrating on designs incorporating large or monolithic sized crystal. A number of companies offer custom-made souvenirs by taking 3D pictures or photos and engraving them into the crystal.

Laser Beam Machining

Laser beam machining is a thermal machining process which uses laser beam to produce heat. In this machining process metal is removed by melting and vaporization of metal particle from surface of work piece. It is a non-conventional machining process in which no tool is used. It is used to machine both metallic and non-metallic material. It is mostly used in cutting and drilling operation.

Principle

The word laser stands for Light amplification by Stimulated Emission of Radiation. When an electron of atom absorbed energy form an external source, the electron which are in its original energy level, jump to a higher energy level. This is not a stable condition of atom so this electron emits the absorbed energy in form of photons and come back to its original state. If an atom which is already at higher energy level absorbs energy, it will emit double energy to return at its original state. The energy emitted by the atom has same frequency and wavelength as the stimulating energy. This is fundamental of laser. When the laser material placed in presence of some other energy source, it absorb energy at some extant and release it when reaches its absorbing limit. This high amplified light beam is called laser.

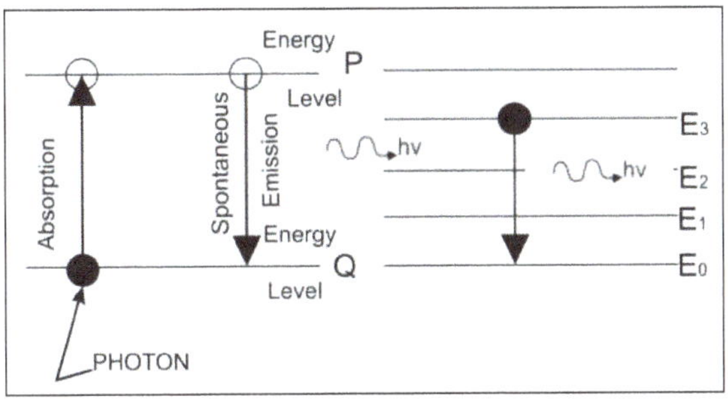

Principle of Laser Beam Welding.

This machining process works on basic property of laser. It uses a laser beam, which is a narrow, monochromatic high intense light which can cut or machine any metal and non-metal. It can use to cut any material irrespective to hardness of work piece. It can also use to cut diamond which is hardest known material on earth.

Equipment's

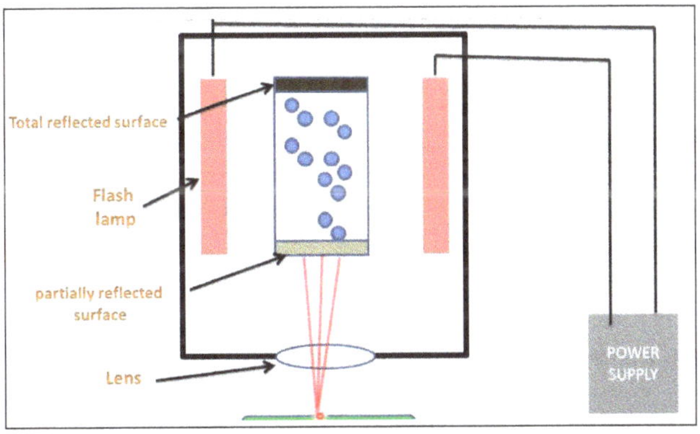

Laser Beam Machining.

Power Supply

It provides the energy for excitation of electron from lower energy level to higher energy level. This gives power to xenon flash lamps, which produce light energy. The laser material are exposed in light energy to keep storing energy.

Laser Discharge Tube

The laser material filled in lased discharge tube. The excitation of electron and come back to its original state process takes place in it. It's one side is partially transparent for laser opening and other side is 100% reflected. It is situated between flash lamp.

Laser Material

There are many different type of laser material available but in later machining mostly CO_2 (Pulsed or continuous waves) and Nd: YAG is used. Carbon die oxide is a laser material that emits light in infrared region. It can provide up to 25 KW power in continuous wave mode. The other one is called Neodymium doped Yttrium Aluminum Garnet is a solid state laser which can delivery light through optical fiber. It can generate about 50 KW power in pulsed mode and 1 KW power in continuous mode.

Focusing Lens

A focusing lens is used in laser machining operation. It is a convex lens which focus is at work piece.

Working

As we know in laser energy is used to remove metal from workpiece. Its process can be summarized as follow:

- First laser material CO_2 or other gases filled into laser discharge tube.
- Now switch on the power supplied which is connected by flash lamp. This lamps produce light energy which used to excite electrons of atom.
- The atoms of laser material absorb energy from the light energy produced by flash lamp. It leads jump of orbital electron of atom form low energy level to high energy level. This is unstable condition of atom.
- This energy initially blind up in laser material. When the atoms absorb sufficient energy it start emit energy continuously. This is high amplified same frequency and same wavelength coherent light.
- This laser light collected by the focus lens and directed toward the work piece.
- Now the laser impinging on work piece start machining process by melting or vaporize material from contact surface.
- This is whole process of laser beam machining.

Application

- Used to drill small hole of diameter about 0.005 mm in refectory and ceramic materials.
- It is used in drilling and cutting for both metals and non-metals.
- It is extensively used in electronic and automotive industries.

- It is mostly used in aerospace industries.
- Used to machine complex profile where machining by tool is not possible.

Advantages and Disadvantages

Advantages

- It can cut all material.
- No tooling cost because no physical tool is required.
- It produces finish part or high surface finish.
- No tool wear because no physical tool is used.
- Micro holes can be drilled accurately.
- Complex shape can be machined easily because laser can be move in any path.
- Very hard material can be cut through laser beam machining.
- High accuracy can be achieved.
- It can be easily automated and flexible.

Disadvantages

- It is uneconomical when high volume of same shape to be cut compare to stamping.
- High capital and maintenance cost.
- It cannot use to produce blind hole.
- Laser can leads to safety hazards.

Laser Ablation

Laser ablation is the process by which layers are removed from solid metals and industrial compounds using a laser beam for ultimate precision. The beam will irradiate the surface, meaning that it has been exposed to radiation.

Laser Ablation Process

Laser ablation works by focusing a laser onto a substrate to a remove material that is on its surface. The amount that is removed depends on the intensity, pulse length, and wavelength of the laser, as well as the material itself. The area absorbs the laser that is being directed on it, thereby breaking down the chemical bonds within the area.

With a low level of laser flux, the material being focused upon is absorbed by the laser's energy and then changes to a gaseous state. With a higher level of laser flux, the material that is being focused upon will usually be converted to plasma. It is possible to undergo the laser ablation process with both a pulsed fiber laser and a continuous wave laser, although the former is the more common method due to the high level of laser intensity.

Laser ablation has many benefits over more traditional methods of processes, such as with thin film removal, whereby alternative solutions have to undergo a multi-step process which is costly, time-consuming and inflexible, as well as having risks for the environment. Laser ablation is a much more efficient, reliable and cost-effective method.

What uses does Laser Ablation have?

Laser ablation has many different uses, particularly as the whole process is undergone with minimal excess heat being transferred to the surrounding area of the material being used. This means that the process will have little to no effect on the parts of the material that you don't want it to, and so laser ablation is great for a wide range of materials, including plastic, metal, ceramic, and even biological tissue. With the advanced nature of fiber lasers, you can focus on a very small, specific spot to produce your desired results.

Laser ablation can also be used to determine the presence and concentration levels of a particular chemical or material on a surface. This is achieved by generating bright plasma on the surface, and then analysing this plasma to see what is present there. This is a much more environmentally-friendly process for determining chemical analysis, as opposed to more traditional methods such as using toxic acid solutions.

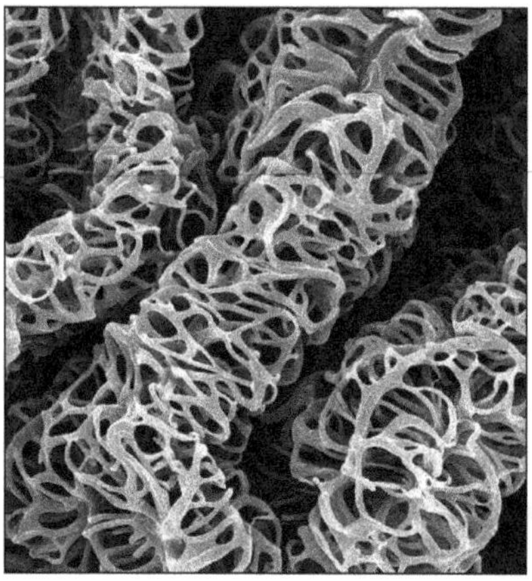

Laser ablation is also useful for viewing the micro-structuring of an object.

Finally, laser ablation can be used to deposit film on a surface, so as to produce things like Nanomaterials or to fabricate superconducting materials. This can also be used for the micro-structuring of an object, whereby the structuring of a larger object can be seen.

Factors affected to Laser Ablation

The success and efficiency of the laser ablation process depend on a number of factors. These are:

Wavelength

The wavelength needs to be selected carefully with a minimum absorption depth. This is so that there is a high energy deposition within a small volume, resulting in a quick and efficient laser ablation.

Pulse Duration

It is better to use shorter pulse durations so as to perform at maximum peak power and to minimise the thermal damage that is inflicted on the surrounding area, although this is already greatly reduced due to the precise nature of the laser ablation process.

Pulse Repetition Rate

The pulse repetition rate needs to be high enough so that the heat caused by laser ablation is maintained and is not left any time to cool. This will result in a more successful ablation as energy waste will be reduced.

Beam Quality

Your laser beam needs to be of sufficient quality to perform the laser ablation process successfully. The quality of the beam will be determined by its brightness, its focus ability, and its homogeneity. The beam size will also need to be controlled so that you don't ablate too large an area.

Application Areas for Laser Ablation

Due to the precise, accurate and environmentally-friendly nature of laser ablation, it is used in many different application areas and industries. Its ability to remove thin film has particular benefits for the electronics and semiconductor industries, and it has also found uses for other industries such as medical, automotive and ship-building. It has understandably become the preferred solution for many manufacturing processes.

Benefits of Laser Ablation

One benefit of laser ablation is that it can be conducted using both a pulsed wave and continuous wave laser, although the former is the primary method as a high level of intensity is needed with the laser beam being used.

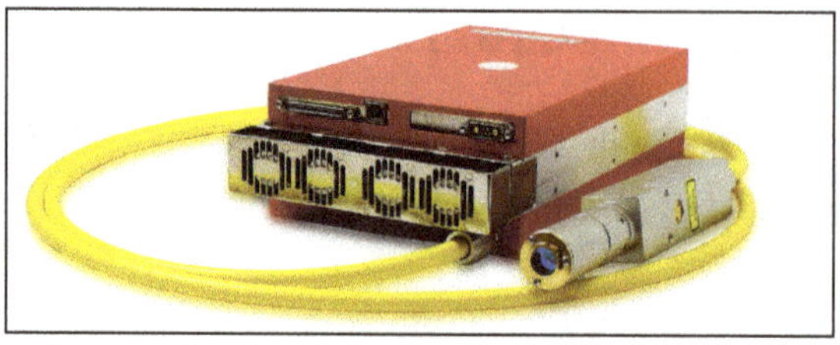

Laser ablation can be completed with both pulsed lasers and continuous wave lasers.

Minimal Heat Transfer

Despite this high level of intensity, very little heat energy is transferred to the area surrounding the spot which you are performing the laser ablation on. This is extremely beneficial in that you can rest easy knowing you won't be causing damage to the rest of the material that you are working with. And, as you can focus on a small spot, you can remove as much or as little material as you like.

Cost-effective

It is also an extremely cost-effective solution for many different processes. For example, with thin film removal, traditional methods require a multi-step approach; having to create a pattern for the area that is to be removed and then using chemicals to complete the task. Laser ablation, on the other hand, is quick, efficient, and highly cost-effective, making it the natural solution to any thin film removal process.

Environmentally Friendly

As well as its cost-effective and low energy waste benefits, laser ablation is a safer and more environmentally-friendly approach as it uses no solvents, it is relatively easy to automate with robots and is much gentler than using more abrasive techniques such as dry-ice blasting. This results in it being the ideal choice for a wide range of materials, including metals, ceramics, and plastics.

The Different uses for Laser Ablation

Not only does laser ablation have many applications for materials in industries such as the electronics sector, for applications such as thin film removal. It is a proven method for treating epilepsy, removing brain tumours, treating spine tumours and even for the resurfacing of human skin. Other industries laser ablation is popular within include Aerospace and Automotive.

Alongside the clear benefits of laser ablation for the removal and treatment of certain conditions, it is also used for:

- Propulsion – Laser ablation can transfer momentum to a surface, which is used to work-harden metal surfaces.

- Manufacturing – Electronic semiconductors and microprocessors are laser ablated as standard now, and it is also used in processes such as the removal of barriers from gas turbine components.
- Science – Laser ablation is regularly used within the scientific field to study the function of nerves and tissue by destroying them.

While its primary function is to remove layers of a material, it is also used to deposit film, which is useful for nanocrystals and nanotubes, for material characterisation, which is the process of determining which materials are present on an object, and for micro-structuring, which is the fabricating of miniature structures of objects.

As can be seen, laser ablation is a multi-functioning, multi-purpose process which can be accurately controlled to provide extremely precise and efficient results. Its ability to adapt to a variety of different industries, whether this is manufacturing or science, makes it extremely useful, particularly as it can be used on everything from metal to human skin.

Applications for Laser Ablation

When lasers were first discovered as a potential use in industries back in the 1960s, they perhaps weren't taken as seriously as they deserved at first. While the scientific community recognised their importance, they didn't really see how lasers could fit into the wider world around them.

It wasn't long until they started to enjoy commercial use, and since then they have grown to be a crucial part of many sectors around the world. From heavy industrial settings in the automotive and aerospace industries, to key areas like helping to save lives and the everyday electronics they use, they aren't only limited to manufacturing plants.

Ablation with a laser itself can be used in three primary ways. The most common is the removal of a surface layer of a material, as we've examined above (e.g. thin film removal). The second is its ability to deposit film onto a surface area. The final use is for determining the presence of materials or chemicals on the surface layer of another material. With multi-functional purposes such as these, it is easier to understand why this has become such a widely used process.

Various Surgeries

Given the fact that it can work with human tissue, it is becoming an important process in the surgical world. By placing a thin laser probe in places such as the brain, it has been used to remove brain tumours and spine tumours. It's even showing promise for the treatment of prostate cancer too.

Alongside the treatment of cancer, it can perform other surgeries too. It is used to treat epilepsy, as well as the cauterising and closing of veins in one's legs. While this may be for cosmetic reasons, it is generally done to alleviate pain and aching. It offers a more precise and safer alternative to many traditional methods, and also has fewer side effects too.

The Resurfacing of Human Skin

Other ways that it is used with biological tissue is for the resurfacing of human skin. Laser ablation is able to resurface with ablation superficial, medium and deep wounds can all be cleaned to great effect.

It is generally used to treat photo damaged skin or scars, but can also be used for facelifts as well. It is a process that can be used on human skin.

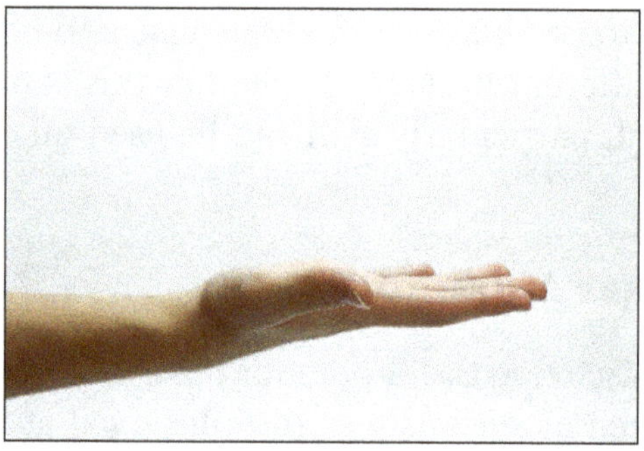

Dentistry

Given the fact that laser ablation can remove layers off a surface of a material while causing minimal damage to the surrounding area, it is widely used by dentists. One of the most common applications in this industry is for removing tooth enamel.

Cleaning Surfaces

Remember that we mentioned that ablation is very similar to laser cleaning? The process is widely used for the cleaning of a number of surfaces. Reasons that this may be done are to remove paint from a surface, cleaning a surface ready for use again (such as baking trays), or to prepare a surface for other types of treatment and process.

Another key application for cleaning is when it is used with metal, helping to easily remove rust from the surface layer.

Coating Surfaces

Earlier we looked at how ablation with a laser has three primary functions, and one of these was to deposit a material onto the surface of another material. This is done by laser ablating one material, and depositing the residue on the other material that you want to work with.

This can be done to coat the surface layer, helping to create a high-quality end-product that can't be easily evaporated. For example, it is regularly undertaken for semiconductors and nanomaterials.

Transferring Momentum

Another use for is for the transferring of momentum. As the process applies a high amount of pressure to the surface that it is working on, it can help to harden this surface. It's extremely similar to (as if you were) hitting the surface with a hammer.

Scientific Uses

The fact that it can determine chemicals and materials present on a surface is highly useful for the scientific community. This is a much more environmentally-friendly way of doing it compared to past methods, where often toxic chemical solutions would have to be used.

It is also used by scientists to research the way in which nerves function, as well as tissues. Most of this research occurs by destroying the tissue.

References

- Gan, E.K.W.; Zheng, H.Y.; Lim, G.C. (7 Dec 2000). Laser drilling of micro-vias in PCB substrates. Proceedings of 3rd Electronics Packaging Technology Conference. IEEE. Doi:10.1109/eptc.2000.906394. ISBN 0-7803-6644-1
- Laser-pointers: rp-photonics.com, Retrieved 22 June, 2019
- "How Fiber Laser Technology Compares to CO2 - Boss Laser Blog". Boss Laser Blog. 2017-05-22. Retrieved 2018-04-24
- What-is-laser-ablation, application-ablation: spilasers.com, Retrieved 26 January, 2019
- Voisey, K.T.; Cheng, C.F.; Clyne, T.W. (2000). "Quantification of Melt Ejection Phenomena During Laser Drilling". MRS Proceedings. San Francisco: Cambridge University Press (CUP). 617. Doi:10.1557/proc-617-j5.6. ISSN 0272-9172
- Benefits-laser-ablation, application-ablation: spilasers.com, Retrieved 27 February, 2019
- Forsman, A; et al. (June 2007). "Superpulse A nanosecond pulse format to improve laser drilling" (PDF). Photonics Spectra. Retrieved June 16, 2014
- Laser-ablation-applications, application-ablation: spilasers.com, Retrieved 28 March, 2019
- Laser-beam-welding: theweldingmaster.com, Retrieved 23 July, 2019
- Laser-beam-welding: theweldingmaster.com, Retrieved 21 March, 2019
- Bovatsek, Jim; Tamhankar, Ashwini; Patel, Rajesh (November 1, 2012). "Ultraviolet lasers: UV lasers improve PCB manufacturing processes". Laser Focus World. Retrieved 20 July 2014.
- Aser-beam-machining-principle-working-equipment-application-advantages-and-disadvantages: mech4study.com, Retrieved 25 August, 2019

PERMISSIONS

All chapters in this book are published with permission under the Creative Commons Attribution Share Alike License or equivalent. Every chapter published in this book has been scrutinized by our experts. Their significance has been extensively debated. The topics covered herein carry significant information for a comprehensive understanding. They may even be implemented as practical applications or may be referred to as a beginning point for further studies.

We would like to thank the editorial team for lending their expertise to make the book truly unique. They have played a crucial role in the development of this book. Without their invaluable contributions this book wouldn't have been possible. They have made vital efforts to compile up to date information on the varied aspects of this subject to make this book a valuable addition to the collection of many professionals and students.

This book was conceptualized with the vision of imparting up-to-date and integrated information in this field. To ensure the same, a matchless editorial board was set up. Every individual on the board went through rigorous rounds of assessment to prove their worth. After which they invested a large part of their time researching and compiling the most relevant data for our readers.

The editorial board has been involved in producing this book since its inception. They have spent rigorous hours researching and exploring the diverse topics which have resulted in the successful publishing of this book. They have passed on their knowledge of decades through this book. To expedite this challenging task, the publisher supported the team at every step. A small team of assistant editors was also appointed to further simplify the editing procedure and attain best results for the readers.

Apart from the editorial board, the designing team has also invested a significant amount of their time in understanding the subject and creating the most relevant covers. They scrutinized every image to scout for the most suitable representation of the subject and create an appropriate cover for the book.

The publishing team has been an ardent support to the editorial, designing and production team. Their endless efforts to recruit the best for this project, has resulted in the accomplishment of this book. They are a veteran in the field of academics and their pool of knowledge is as vast as their experience in printing. Their expertise and guidance has proved useful at every step. Their uncompromising quality standards have made this book an exceptional effort. Their encouragement from time to time has been an inspiration for everyone.

The publisher and the editorial board hope that this book will prove to be a valuable piece of knowledge for students, practitioners and scholars across the globe.

INDEX

www.ingramcontent.com/pod-product-compliance
Lightning Source LLC
LaVergne TN
LVHW080632210926
848485LV00003B/250

* 9 7 8 1 6 3 2 3 8 8 5 8 2 *